Lecture Notes in Computer

Commenced Publication in 1973
Founding and Former Series Editors:
Gerhard Goos, Juris Hartmanis, and Jan van Leeuwen

Editorial Board

David Hutchison
Lancaster University, UK

Takeo Kanade
Carnegie Mellon University, Pittsburgh, PA, USA

Josef Kittler
University of Surrey, Guildford, UK

Jon M. Kleinberg
Cornell University, Ithaca, NY, USA

Friedemann Mattern
ETH Zurich, Switzerland

John C. Mitchell
Stanford University, CA, USA

Moni Naor
Weizmann Institute of Science, Rehovot, Israel

Oscar Nierstrasz
University of Bern, Switzerland

C. Pandu Rangan
Indian Institute of Technology, Madras, India

Bernhard Steffen
University of Dortmund, Germany

Madhu Sudan
Massachusetts Institute of Technology, MA, USA

Demetri Terzopoulos
University of California, Los Angeles, CA, USA

Doug Tygar
University of California, Berkeley, CA, USA

Moshe Y. Vardi
Rice University, Houston, TX, USA

Gerhard Weikum
Max-Planck Institute of Computer Science, Saarbruecken, Germany

Yannis Avrithis   Yiannis Kompatsiaris
Steffen Staab   Noel E. O'Connor (Eds.)

Semantic
Multimedia

First International Conference on Semantic
and Digital Media Technologies, SAMT 2006
Athens, Greece, December 6-8, 2006
Proceedings

 Springer

Volume Editors

Yannis Avrithis
Image, Video & Multimedia Systems Laboratory
School of Electrical and Computer Engineering
National Technical University of Athens
9, Iroon, Polytechniou Str., 15773 Zographou, Athens, Greece
E-mail: iavr@image.ntua.gr

Yiannis Kompatsiaris
Informatics and Telematics Institute
Centre for Research and Technology Hellas
6th Km Thessaloniki-Thermi Road, 57001 Thermi-Thessaloniki, Greece
E-mail: ikom@iti.gr

Steffen Staab
University of Koblenz-Landau
Universitaetsstrasse 1, 56016 Koblenz, Germany
E-mail: staab@uni-koblenz.de

Noel E. O'Connor
Centre for Digital Video Processing
Dublin City University
Collins Avenue, Dublin 9, Ireland
E-mail: oconnorn@eeng.dcu.ie

Library of Congress Control Number: 2006936895

CR Subject Classification (1998): H.5.1, H.4, H.5, I.2.10, I.4, I.7, C.2

LNCS Sublibrary: SL 3 – Information Systems and Application, incl. Internet/Web
and HCI

ISSN   0302-9743
ISBN-10   3-540-49335-2 Springer Berlin Heidelberg New York
ISBN-13   978-3-540-49335-8 Springer Berlin Heidelberg New York

This work is subject to copyright. All rights are reserved, whether the whole or part of the material is concerned, specifically the rights of translation, reprinting, re-use of illustrations, recitation, broadcasting, reproduction on microfilms or in any other way, and storage in data banks. Duplication of this publication or parts thereof is permitted only under the provisions of the German Copyright Law of September 9, 1965, in its current version, and permission for use must always be obtained from Springer. Violations are liable to prosecution under the German Copyright Law.

Springer is a part of Springer Science+Business Media

springer.com

© Springer-Verlag Berlin Heidelberg 2006
Printed in Germany

Typesetting: Camera-ready by author, data conversion by Scientific Publishing Services, Chennai, India
Printed on acid-free paper   SPIN: 11930334   06/3142   5 4 3 2 1 0

Preface

We are delighted to welcome you to the proceedings of the 1st International Conference on Semantic and Digital Media Technologies held in Athens.

SAMT 2006 aims to narrow the large disparity between the low-level descriptors that can be computed automatically from multimedia content and the richness and subjectivity of semantics in user queries and human interpretations of audiovisual media — The Semantic Gap. SAMT started out as two workshops, EWIMT 2004 and EWIMT 2005, that quickly achieved success in attracting high-quality papers from across Europe and beyond. This year EWIMT turned into the full-fledged conference SAMT, bringing together forums, projects, institutions and individuals investigating the integration of knowledge, semantics and low-level multimedia processing, and linking them with industrial engineers who exploit the underlying emerging technology.

In total, 68 papers were submitted to the SAMT 2006 conference and each was reviewed by at least two independent reviewers. We are grateful to the members of the Technical Program Committee who completed these reviews and allowed us to put together a very strong technical program of 17 papers. The selection process was very competitive with only 25% of papers being selected for oral presentation. The program also included two invited keynote talks from Alan Smeaton and Guus Schreiber, and we are very grateful to them for their insightful presentations.

We thank all our colleagues in the Organization Committee who helped us tremendously with putting together an interesting and rewarding number of events including four workshops, three tutorials, three special sessions, poster, demo and projects sessions, the SAMT 2006 industry day, and two invited talks from Roberto Cencioni and Luis Rodriguez-Rosello featuring the launch of the 7th Framework ICT Program. Our Steering Committee was of great assistance in all critical decisions, and the staff of the Image, Video and Multimedia Systems Laboratory of the National Technical University of Athens provided invaluable help in making this conference happen. Special thanks go to Marios Phinikettos and Themis Zervou for the tremendous amount of work they put into supporting SAMT 2006.

The SAMT 2006 conference was held in cooperation with the European Commission, EURASIP, COST 292 and the IET, and we are grateful to these organizations for promoting the event. We are also indebted to K-Space Network of Excellence, of which SAMT was envisioned as the flagship event, and all sponsors for their contributions and financial support.

December 2006

Y. Avrithis
Y. Kompatsiaris
S. Staab
N.E. O'Connor

International Conference on Semantic and Digital Media Technologies

Organizing Committee

General Chairs
: Yannis Avrithis
  National Technical University of Athens, Greece
  Yiannis Kompatsiaris
  Informatics and Telematics Institute, Greece

Program Chairs
: Noel O'Connor
  Dublin City University, Ireland
  Steffen Staab
  University of Koblenz-Landau, Germany

Special Sessions
: José Martinez
  Universidad Autonoma de Madrid, Spain

Tutorials and Workshops
: Raphaël Troncy
  CWI, Netherlands
  Vassilis Tzouvaras
  National Technical University of Athens, Greece

EU Liaison
: Paola Hobson
  Motorola Labs, UK

Programme Committee

Bruno Bachimont, INA, France
Wolf-Tilo Balke, University of Hannover, Germany
Jenny Benois-Pineau, University of Bordeaux, France
Jesús Bescós, GTI-UAM, Spain
Nozha Boujemaa, INRIA, France
Patrick Bouthemy, IRISA, France
Pablo Castells, NETS-UAM, Spain
Andrea Cavallaro, Queen Mary University of London, UK
Stavros Christodoulakis, Technical University of Crete, Greece
Fabio Ciravegna, University of Sheffield, UK
Thierry Declerck, DFKI, Germany
Anastasios Delopoulos, Aristotle University of Thessaloniki, Greece
Edward Delp, Purdue University, USA
Martin Dzbor, Knowledge Media Institute, UK
Touradj Ebrahimi, Swiss Federal Institute of Technology, Switzerland
Jerome Euzenat, INRIA, France

Borko Furht, Florida Atlantic University, USA
Moncef Gabbouj, Tampere University of Technology, Finland
Christophe Garcia, France Telecom, France
William I. Grosky, University of Michigan, USA
Werner Haas, Joanneum Research, Austria
Christian Halaschek-Wiener, University of Maryland, USA
Siegfried Handschuh, DERI Galway, Ireland
Alan Hanjalic, TU Delft, Netherlands
Lynda Hardman, CWI, Netherlands
Andreas Hotho, University of Kassel, Germany
Jane Hunter, University of Queensland, Australia
Antoine Isaac, Vrije Universiteit Amsterdam, Netherlands
Ebroul Izquierdo, Queen Mary University of London, UK
Franciska de Jong, University of Twente, Netherlands
Joemon Jose, University of Glasgow, UK
Aggelos K. Katsaggelos, Northwestern University, USA
Hyoung Joong Kim, Kangwon National University, Korea
Joachim Koehler, Fraunhofer IMK, Germany
Stefanos Kollias, National Technical University of Athens, Greece
Paul Lewis, University of Southampton, UK
Petros Maragos, National Technical University of Athens, Greece
Ferran Marques, Technical University of Catalonia, Spain
Jose Martinez, GTI-UAM, Spain
Adrian Matellanes, Motorola Labs, UK
Bernard Merialdo, EURECOM, France
Frank Nack, V2_ Institute for the Unstable Media and CWI, Netherlands
Jacco van Ossenbruggen, CWI, Netherlands
Jeff Pan, University of Manchester, UK
Thrasyvoulos Pappas, Northwestern University, USA
Ioannis Pitas, Aristotle University of Thessaloniki, Greece
Dietrich Paulus, University of Koblenz, Germany
Eric Pauwels, CWI, Netherlands
Dennis Quan, IBM, USA
Keith van Rijsbergen, University of Glasgow, UK
Andrew Salway, University of Surrey, UK
Mark Sandler, Queen Mary, University of London, UK
Simone Santini, University of California, San Diego, USA
Guus Schreiber, Vrije Universiteit Amsterdam, Netherlands
Timothy Shih, Tam Kang University, Tawain
Sergej Sizov, University of Koblenz, Germany
Alan Smeaton, Dublin City University, Ireland
John R. Smith, IBM Research, USA
Giorgos Stamou, National Technical University of Athens, Greece
Fred S. Stentiford, University College London, UK
Michael Strintzis, ITI, Greece

Rudi Studer, University of Karlsruhe, Germany
Vojtech Svatek, UEP, Czech Republic
Murat Tekalp, University of Rochester, USA
Raphaël Troncy, CWI, Netherlands
Paulo Villegas, Telefonica I+D, Spain
Gerhard Widmer, University of Linz, Austria
Li-Qun Xu, British Telecom, UK

Sponsors

SAMT 2006 was organized by the Image, Video and Multimedia Systems Laboratory at the National Technical University of Athens. The event was co-sponsored by:

Publicity Sponsors

 Information Society Technologies (IST), European Commission

 European Association for Signal, Speech and Image Processing (EURASIP)

 European Cooperation in the field of Scientific and Technical Research, COST 292 Action — Semantic Multimodal Analysis and Digital Media

 Visual Information Engineering Professional Network (VIE PN), Institution of Engineering and Technology (IET)

Emerald Sponsor

 K-Space Network of Excellence — Knowledge Space of Semantics Inference for Automatic Annotation and Retrieval of Multimedia Content

Gold Sponsor

 Muscle Network of Excellence — Multimedia Understanding through Semantics, Computation and Learning

Silver Sponsors

 aceMedia Integrated Project — Integrating Knowledge, Semantics and Content for User-Centred Intelligent Media Services

MOTOROLA Motorola UK Ltd

Industry Day Sponsors

 Athens Technology Center (ATC)

 Deutsche Welle

 MESH Integrated Project — Multimedia Semantic Syndication for Enhanced News Services

Best Paper Award Sponsor

X-Media Integrated Project — Large Scale Knowledge Sharing and Reuse Across Media

Table of Contents

Content vs. Context for Multimedia Semantics: The Case of SenseCam Image Structuring

Invited Keynote Paper

Alan F. Smeaton

Adaptive Information Cluster
& Centre For Digital Video Processing,
Dublin City University,
Ireland
Alan.Smeaton@dcu.ie

Abstract. Much of the current work on determining multimedia semantics from multimedia artifacts is based around using either context, or using content. When leveraged thoroughly these can independently provide content description which is used in building content-based applications. However, there are few cases where multimedia semantics are determined based on an *integrated* analysis of content and context. In this keynote talk we present one such example system in which we use an integrated combination of the two to automatically structure large collections of images taken by a SenseCam, a device from Microsoft Research which passively records a person's daily activities. This paper describes the post-processing we perform on SenseCam images in order to present a structured, organised visualisation of the highlights of each of the wearer's days.

1 Introduction

When we think of multimedia information retrieval and multimedia semantics we tend to think of fairly standard multimedia artifacts such as still images, music, video and maybe 3D objects. When we think of how to determine the semantic content of such multimedia artifacts we do so because we want to perform a variety of content-based operations on such information including browsing, searching, summarisation, linking, etc. And finally, when we look at *how* we might determine semantics of multimedia objects we find that there are generally two approaches, namely:

1. use the context of the objects such as information gathered at the time of object creation or capture, to help determine some content features;
2. extract information directly from within the content of the objects in order to determine some content aspects.

Trying to determine and then usefully use a user's *context* is a fairly hot topic in information retrieval at the moment with lots of attempts to capture and then

Y. Avrithis et al. (Eds.): SAMT 2006, LNCS 4306, pp. 1–10, 2006.
© Springer-Verlag Berlin Heidelberg 2006

apply such context in retrieval [11]. Determining a document or a multimedia object's context has also been explored for a long time and this forms the basis for many current systems for multimedia object management. For example such basic metadata as date and time of creation form the essential content representation for many tools which manage personal photos. Examples of such popular photoware includes Photoshop Album [2], PhotoFinder [16], ACDSee [1], Picasa [9] and others. Other metadata created at the time of photo capture such information as shutter speed and lens aperture, whether a flash was used or not can also be used to support automatic grouping of photos [14]. Finally, there are emerging online photoware systems such as Flickr [5] and Yahoo 360 [19] which support user-supplied context information to help with photo organisation.

What all these applications have in common, apart from the fact that they are all used to manage personal photos, is that they all use semantic information to describe multimedia objects (photos) which are derived from the context of the photo ... either directly from the capture process, or provided by an end-user afterwards.

To complement semantics derived from context we also use semantics derived from content in helping to manage our multimedia objects. Returning to the example of personal photos, this corresponds to extracting features directly from the image contents. An example system which does this is MediAssist which automatically determines whether a picture was taken indoors or outdoors, whether it is of a built or of a natural environment, whether a picture contains faces and if so whether those faces are faces of known individuals [14]. While this is a limited set of descriptive semantics, automatically determining the presence or absence of a larger number of medium and high level semantic features in visual media is notoriously difficult as is shown repeatedly in the TRECVid benchmarking evaluation campaign [17].

Once we have determined some level of semantic representation for multimedia objects we can then use these for content-based operations such as retrieval and we find that those derived from content and from context are almost always used either independently of each other or collaboratively with each other, but rarely are they truly integrated with each other. In other words, because these semantics are derived from different primary sources they maintain and retain their differing heritages when they are used subsequently.

To illustrate this let us examine the different ways in which video shots can be retrieved. In [18] we presented a classification of five different experimental approaches to video shot retrieval, namely:

1. Use metadata determined at the time of video capture/creation to access video by date, time, title, genre, actors, popularity rating, etc. as in [12];
2. Use one or more example query images to match against shot keyframes using whole-image matching approaches based on colour, texture or edges, as shown by many systems in [17];
3. Use text queries to match against text derived from transcriptions of the spoken dialogue of text determined from video OCR, also as shown by many systems in [15];

4. Use video objects, semi-automatically determined from shot keyframes and from user query images, and match these video objects based on shape, colour and/or texture, also as shown by many systems in [17];

5. Use the presence or absence of semantic video features such as indoor, outdoor, beach, sky, boats, motor vehicles, certain named persons, etc. to narrow the scope of shot retrieval to only those shots likely to contain such features;

Many systems have been developed to support video shot retrieval using one, two or perhaps three of the above but none have been developed to support all of them and for those that support multiple modalities for shot retrieval, the user is normally left with responsibility for combining and integrating them.

In this paper we argue for a more integrated approach to using semantic features determined from content and from context, and we illustrate what is possible with a novel application based around sets of images taken with a Sense-Cam. In the next section we introduce the SenseCam and its possible range of applications and in section 3 we present a summary of our work on structuring SenseCam images based on an integrated combination of content and context features. Section 4 concludes the paper.

2   The SenseCam

A SenseCam is a device developed by Microsoft Research in Cambridge, UK, for recording visual images of a wearer's day. It passively captures images through a fisheye lens and stores them on-board for subsequent download to a personal computer [8]. In addition to being a camera, a SenseCam also has other sensors including a light meter, a passive infra-red sensor and a 3-axis accelerometer and sensor readings from all these devices are also stored for later download. However, in addition to recording some elements of the SenseCam environment, the additional sensors are also used in a semi-intelligent way to trigger when photos are to be taken. For example when a person walks in front of the wearer this can be picked up by the passive infra-red sensor to trigger a photo to be taken. Similarly, when the user moves by standing up, or moves from indoor to outdoor or vice-versa, these are picked up by the accelerometer and light level sensors respectively and also trigger taking of photos. As a default, without an explicit triggering from the sensors, or from a user-controlled button on the SenseCam, the device will take a new photo every 45 seconds anyway. In this way a typical day can have up to 3,000 photos taken, which could add up to almost a million images in a year. A SenseCam being worn around a wearer's neck is shown in Figure 1 and a set of sample images taken from the author's use of a SenseCam is shown in Table 1.

The SenseCam device has been used extensively in the MyLifeBits project at Microsoft Research [6], [7] as well as being used in other, exploratory projects at Microsoft Research in Cambridge [10]. Like many other sensor devices, the SenseCam is great at capturing raw data, up to a million images per year for each user, and the main challenge is to effectively manage this huge volume of personal data. This requires automatic analysis and structuring in order to

Table 1. Sample SenseCam images

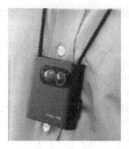

Fig. 1. A SenseCam worn around the neck

impose some organisation on the raw images and this is the challenge we address as we seek to determine semantics for these multimedia objects and to use both context (date, time, sensor readings) and content (image processing) to achieve this. In the next section we describe how we do this.

3   Structuring SenseCam Images

Effectively managing a growing collection of up to 3,000 images taken per day is physically impossible unless the images are structured in some way. Within our daily lives, our activities can be broken down into "events" corresponding to things like having breakfast, walking to the bus stop, travelling to work on the bus, walking to our workplace, making coffee as soon as we arrive at work, sitting at our desk reading email, drinking coffee and starting to write a report, breaking to have a short meeting with colleagues, going to the canteen to have a morning coffee break, returning to work at the desk, having lunch with a group of friends, back to the desk in the afternoon and finishing work with a one-on-one meeting with the boss, getting the bus to the gym, having a workout there, going to a movie, taking the bus home, making and eating dinner, watching TV, and finally going to bed.

While we could argue about the definition of an event, whether travel to-from work is one event or divided into walking to the bus stop, travelling on the bus and waking to work which are each events, in general we can say that each of the above is characterised by being visually different form the preceeding and succeeding events. What the user (and SenseCam wearer) sees will be different for each event because the location will change or the people present will change. In theory, such changes in location are detectable through processing the sets of images taken during each event. In a way this is analogous to the task of shot boundary detection in video where we also wish to find the boundary between different shots by comparing images, but in the case of SenseCam event segmentation the task is more difficult because adjacent images may be quite different from each other but still part of the same event. These image differences will be caused by the user turning around towards/away from a window or light source or facing in a different direction, looking at different people, or a different part of the same room. However the *set* of images constituting an event will be globally similar to each other. In contrast, adjacent images in video will only have small differences, unless there is a photo flash or some very rapid camera and/or object movement.

In work reported elsewhere we have addressed the problem of event segmentation by comparing temporally adjacent and temporally nearby SenseCam images using conventional low-level image features like colour and texture [3], as well as spatiograms [4], and our results on this to date indicate that using image processing techniques alone we can achieve useful results. When we then incorporate evidence for event boundaries taken from other SenseCam sensor readings and even from detection of local Bluetooth devices such as people's mobile phones [13] then the reliability of event detection improves further.

In our work to date we have found that SenseCam "events" can contain anything from some tens of images to several hundred, depending on the activity taking place as well as the duration of the event. Once events have been detected then we can then further structure a user's SenseCam images by manipulating and reasoning about events themselves. A schematic overview of how we process SenseCam images is shown in Figure 2.

In order to manipulate SenseCam events we need some representation for the event itself so we compute a virtual image as the average of all the SenseCam images within an event. This is a crude first approximation and could be refined by detecting outlier SenseCam images within an event and removing them, or

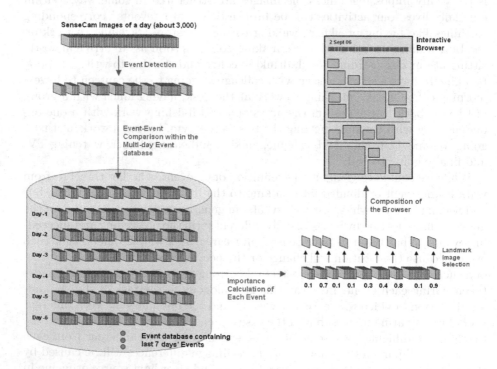

Fig. 2. Schematic for processing SenseCam images

removing or down-weighting SenseCam images towards the beginning and end of detected events as they are more likely to be close to event transitions, which will probably involve the user moving location and this will generate SenseCam images which are not really part of the preceeding or succeeding events. However investigating this aspect is part of our future work. In fact to reduce processing time the event representative is generated as part of the event detection process, so there is little overhead in computing this. Once the virtual representative image from an event is computed we then locate the actual SenseCam image which is visually closest to the virtual centroid and we term that a "landmark" image. The reason for doing this is to use an actual SenseCam image as a representative for presentation of the event. In future work we would like this to be the SenseCam image which has the greatest number of faces present, but for now we base our landmark detection on selection of the image most similar to the virtual average of those in the event.

When a day's SenseCam images are uploaded and the virtual representative for each detected event is available we then add it to a database of event representations. Our task now is to determine which of the day's events are more important than the others. For example, having breakfast, travelling to/from work, having coffee with the same colleagues and working at the same desk are all regular events which happen daily and are not very different from one day to the next, even visually, yet going to the movies, visiting the gym or having lunch in a different restaurant or with different people will all be unusual events for this wearer's lifestyle.

We determine an event's importance by comparing the visual representatives for each event over a fixed 7-day window and examining an event's duration. Basically, if an event is unusual in terms of a given week's activities it will not appear to have any visually similar events or a similar duration and it will then be assigned a high importance or novelty rating. On the other hand if an event is one of a series of regular and repeating events during that 7-day period it will have many similar events, both visually and perhaps in terms of duration also. This is quite an heuristic step and could be refined by considering the time of day for example, but as with landmark detection, using these event features is sufficient for now and a possible topic for future work.

Finally, once a day's SenseCam images have been segmented into events with landmark images and importance ratings determined automatically we can present the day's activities in the browser shown in Figure 3. This browser configuration lays out landmark images from the most important or highly novel events from each day with the size of the landmark image being indicative of the importance rating of the event. In this way the unusual activities are highlighted by being bigger yet the complete set of a day's activities are shown. In Figure 3 we can see that the most unusual events for that day – 31 May 2006 – appear to be the wearer drinking beer with a friend (2nd last landmark image on bottom row) and having meetings with 2 different colleagues as shown in Rows 1 and 2. A timeline bar on the top of the browser indicates the ranges of times during the day when the SenseCam was recording images and also indicates the

Fig. 3. Interface for reviewing a single day's SenseCam images

sizes and relative durations of segmented events. When the user mouses over an event landmark, all the images from that landmark play within the frame of that landmark like a video playback.

Using this browser the wearer can get a complete picture of the day's activities with the set of 2,890 images in the case of Figure 3 being easily navigable because of the way they are structured.

In summary, the processing on a single day's SenseCam images proceeds as follows:

1. Segment the set of images into events using low-level image features and spatiograms, combined with temporal ordering of the images;
2. Generate a virtual event representative as the average of all images in the event;
3. Identify the landmark image for each event as the SenseCam image most visually close to the virtual event representative;
4. Assign each event an importance or novelty rating based on comparing the visual representatives for each event over a fixed 7-day window and examining an event's duration;

On closer examination of the different steps in this process we can see that almost all involve operating on content description derived from both content, and context without any differentiation as to whither the source of that content description. So in this example we make no distinction between content and context in deriving content description and this integration of the sources is to everyone's advantage.

4   Conclusions

In this paper we have examined the sources of information from which multimedia semantics can be derived and categorised them into either content-based or context-based. We have also argued for a more *integrated* approach to determining multimedia semantics where the heritage or origin of the information, whether derived from content or from context, is ignored. To illustrate this we have presented an overview of a complex tool we have developed which ingests a set of several thousands of SenseCam images per day, as a summary of the wearer's daily activities. The interesting aspect of this tool, and the analysis of information gathered by the wearer of the SenseCam, is that the analysis is performed on a combination of context and content based information, with no distinction made between the two sources. This, we believe, is a model of where multimedia semantics should be derived for other applications.

Acknowledgements

This work is partly supported by Science Foundation Ireland under grant number 03/IN.3/I361 with support from Microsoft Research. The work described here is the result of the collaborative research activities of the author and his colleagues, postdoctoral researchers and students.

10   A.F. Smeaton

References

1. ACDSee. Available at http://www.acdsee-guide.com/ (last visited september 2006).
2. Adobe Photoshop Album. Available at http://www.adobe.com/products/-photoshopalbum/ (last visited September 2006).
3. M. Blighe, H. Le Borgne, N. E. O'Connor, A. F. Smeaton, and G. J. F. Jones. Exploiting context information to aid landmark detection in SenseCam images. In *2nd International Workshop on Exploiting Context Histories in Smart Environments (ECHISE)*, Irvine, Calif., USA, September 2006.
4. C. Ó. Conaire, N. E. O'Connor, A. F. Smeaton, and G. J. F. Jones. Organising a Daily Visual Diary Using Multi-Feature Clustering, 2006. submitted for publication.
5. Flickr. Available at http://www.flickr.com/ (last visited September 2006).
6. J. Gemmell, A. Aris, and R. Lueder. Telling Stories with MyLifeBits. In *ICME '05: IEEE International Conference on Multimedia and Expo*, 2005.
7. J. Gemmell, G. Bell, R. Lueder, S. Drucker, and C. Wong. MyLifeBites: Fulfilling the Memex vision. In *Proceedings of ACM Multimedia*, December 2002.
8. J. Gemmell, L. Williams, K. Wood, R. Lueder, and G. Bell. Passive capture and ensuing issues for a personal lifetime store. In *CARPE'04: Proceedings of the the 1st ACM workshop on Continuous archival and retrieval of personal experiences*, pages 48–55, New York, NY, USA, 2004. ACM Press.
9. Google Picasa. Available at http://picasa.google.com/ (last visited september 2006).
10. S. Hodges, L. Williams, E. Berry, S. Izadi, J. Srinivasan, A. Butler, G. Smyth, N. Kapur, and K. Wood. SenseCam: a Retrospective Memory Aid. In *UBICOMP 2006: The 8th International Conference on Ubiquitous Computing*, September 2006.
11. P. Ingwersen and K. Järvelin. *The Turn: Integration of Information Seeking and Retrieval in Context*. Springer: the Kluwer International Series on Information Retrieval, 2005.
12. Internet Archive: Moving Image Archive. Available at http://www.archive.org/details/movies (last visited september 2006).
13. B. Lavelle. SenseCam Social Landmark Detection using Bluetooth, 2006. M.Sc. in Software Engineering Practicum Report, Dublin City University.
14. N. O'Hare, H. Lee, S. Cooray, C. Gurrin, G. J. Jones, J. Malobabic, N. E. O'Connor, A. F. Smeaton, and B. Uscilowski. MediAssist: Using content-based analysis and context to manage personal photo collections. In *CIVR2006 - 5th International Conference on Image and Video Retrieval*, 2006.
15. S. Sav, G. J. Jones, H. Lee, N. E. O'Connor, and A. F. Smeaton. *Interactive Experiments in Object-Based Retrieval*, volume 4071 / 2006, pages 1–10. Springer, Berlin/Heidelberg, Germany, 2006.
16. B. Shneiderman, H. Kang, B. Kules, C. Plaisant, A. Rose, and R. Rucheir. A photo history of SIGCHI: evolution of design from personal to public. *interactions*, 9(3):17–23, 2002.
17. A. F. Smeaton. Large scale evaluations of multimedia information retrieval: The TRECVid experience. In W.-K. L. et al., editor, *CIVR 2005 - International Conference on Image and Video Retrieval*, volume LNCS 3569, pages 11–17, Singapore, July 2005. Springer.
18. A. F. Smeaton. Techniques Used and Open Challenges to the Analysis, Indexing and Retrieval of Digital Video. *Information Systems*, 2006. (in press).
19. Yahoo ! 360. Available at http://360.yahoo.com/ (last visited September 2006).

A Scalable Framework for Multimedia Knowledge Management

Yves Raimond, Samer A. Abdallah, Mark Sandler, and Mounia Lalmas

Centre for Digital Music, Queen Mary, University of London
{yves.raimond, samer.abdallah, mark.sandler}@elec.qmul.ac.uk
Department of Computer Science, Queen Mary, University of London
mounia@dcs.qmul.ac.uk

Abstract. In this paper, we describe a knowledge management framework that addresses the needs of multimedia analysis projects and provides a basis for information retrieval systems. The framework uses *Semantic Web* technologies to provide a shared knowledge environment, and active *Knowledge Machines*, wrapping multimedia processing tools, to exploit and/or export knowledge to this environment. This framework is able to handle a wide range of use cases, from an *enhanced workspace* for researchers to end-user *information access*. As an illustration of how the proposed framework can be used, we present a case study of music analysis.

1   Introduction

Information management is becoming an increasingly important part of multimedia related technologies, ranging from the management of personal collections through to the construction of large 'semantic' databases intended to support complex queries. One of the key problems is the current gap between the development of stand-alone multimedia processing algorithms (such as feature extraction, or compression) and knowledge management technologies. The aim of our work is to provide a framework that is able to bridge this gap, by integrating the multimedia processing algorithms in an information management system, which can then be usable by different entities in different places. We also want to provide a way to semantically describe these algorithms in order to *automatically* use them. For example, we might want to dynamically compute the segmentation of a sequence of a football match in order to answer a query like 'give me all the sequences corresponding to a corner'.

In order to achieve this goal, we introduce several concepts. A *Knowledge Machine* aims to help people developing, encapsulating or testing multimedia processing algorithms, by designing a *semantic workspace*. Instances of such machines interact with a set of *end-points*, which are entry points to a shared knowledge environment, which is itself based upon *Semantic Web* technologies. This interaction can be either to request knowledge from this environment (in order to test algorithms, or to use external information *in* the algorithm) or to export new knowledge onto it (to make new results publicly available). An

Y. Avrithis et al. (Eds.): SAMT 2006, LNCS 4306, pp. 11–25, 2006.
© Springer-Verlag Berlin Heidelberg 2006

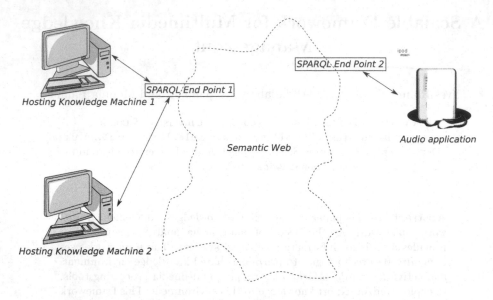

Semantic Web

Hosting Knowledge Machine 1

Hosting Knowledge Machine 2

Audio application

Fig. 1. An overview of the framework

end-point can also have a *planning* [1] role, as part of the knowledge environment describes these *Knowledge Machines*, and the effect of using some of their algorithms. Moreover, these *end-points* can be used by other entities (such as a personal audio player), which may benefit from some information potentially held by this knowledge environment. A simplified overview of the system is given in fig. 1.

The proposed framework is the first step to bridge the above mentioned gap. Indeed, the algorithm is entirely handled from its implementation process to its impact on a shared knowledge environment, which can either be used by other researchers or by information access systems. Moreover, new knowledge can be dynamically added (either from 'ground truth' sources or from Knowledge Machines plugged onto the knowledge environment), as well as new Knowledge Machines. This *web* approach allows the creation of a scalable knowledge management system.

In § 2 we will describe the structure behind the Knowledge Machines. We will then focus on the shared knowledge environment in § 3, and how it deals with instances of Knowledge Machines. Next, we will focus on two completely different use cases of the system in § 4, in order to give an idea of the wide range of possibilities this framework brings. Finally, § 5 will present a case study on knowledge management for music analysis.

2   Knowledge Machines

In this section, we describe the Knowledge Machine architecture (described in greater detail but in a more specific context in [2]), which is able to design a workspace for handling and/or developing multimedia processing algorithms. With complex algorithms, there are often many shared steps of computation, multiple computation strategies, and many free parameters that can be varied to tune performance. This can result in a very large amount of final and intermediate results, which needs to be managed in order to be used effectively. fig. 4 gives a global view of the Knowledge Machines architecture, built around a unique knowledge representation framework (see § 2.1), a computational engine (see § 2.2) and *tabling* of results (see § 2.3).

2.1   Knowledge Representation for Data Analysis

In this section, we will describe the main approach for storing the results of the different computations, which can occur while working on a set of multimedia processing algorithms. Its limitation is what led us to using another approach, which will also be described in this section, using *predicate calculus* for knowledge representation.

The Dictionary Approach. The resulting data is often managed as a dictionary of key-value pairs—this may take the form of named variables in a Matlab workspace, files in a directory, or files in a directory tree (in which case the keys would have an hierarchical structure). This can lead to a situation in which, after a Matlab session for example, one is left with a workspace full of objects but no idea how each one was computed, other than, perhaps, cryptic clues in the form of the variable names one has chosen.

The semantic content of these data is intimately tied to knowledge about which function computed which result using what parameters, and so one might attempt to remedy the problem by using increasingly elaborate naming schemes, encoding information about the functions and parameters into the keys. This is a step toward a relational structure where such information can be represented explicitly and in a consistent way.

Relational and Logical Data Models. We now focus on a relational data model [3], where different relations are used to model the connections between parameters, source data, intermediate data and results. Each tuple in these relations represents a proposition, such as '*this* spectrogram was computed from *this* signal using *these* parameters' (see fig. 2). From here, it is a small step to go beyond a relational model to a deductive model, where logical predicates constitute the basic representational tool, and information can be represented either as facts or as composite formulæ involving the logical connectives *if*, $\exists$ (*exists*), $\forall$ (*for all*), $\vee$ (*or*), $\wedge$ (*and*), $\neg$ (*not*) and $\equiv$ (*equivalent to*) (see [2] for a short review of predicate calculus for knowledge representation).

For example, in this model, the previous proposition could be expressed using this predicate:

$$spectogram(DigitalSignal, FrameSize, HopSize, Spectrogram)$$

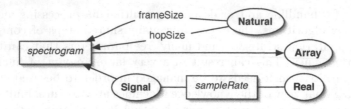

Fig. 2. The relations involved in defining a spectrogram

In addition, we could imagine the following for the digital representation of a signal:

digitalsignal(*DigitalSignal*, *SampleRate*) *if*

∃*ContinuousSignal. sampling*(*ContinuousSignal*, *DigitalSignal*, *SampleRate*)

2.2 Evaluation Engine

The computation-management facet of the Knowledge Machines is handled through calls to an external *evaluation engine*. The latter is used to reduce a given expression to some canonical form. For example, a real-valued expression involving mathematical functions and arithmetic operators would be reduced to the floating-point representation of the result. Standard Prolog itself provides such a facility through the is operator. By using an interface to an interpreted language processor, such as Matlab , a much richer class of expressions can be evaluated, involving complex numbers, arrays, structures, and the entire library of Matlab functions available in the system.

For example, if we define the operator === as evaluating terms representing Matlab expressions, we can define (in terms of predicate calculus) a matrix multiplication like this:

$$mtimes(A, B, C) \; if \; C{===}A * B$$

We can now build composite formulæ involving the predicate mtimes.

Interpreters for different expression languages could be added, provided that a Prolog representation of the target language can be designed.

2.3 Function Tabling

To keep track of computed data, we consider tabling of such logical predicates. Some predicates, when used in a certain way (in a particular mode), can be considered as 'functional'—one possible resulting tuple given a set of inputs (such as mtimes, when used with the first and the second argument bound, and the third one unbound). If we store the tuples generated by the functional predicates, then we can save ourselves some evaluations, because of this functional mode. Moreover, the function tabling mechanism allows us to access functional predicates in *all* modes. Given the produced data, we can obtain back to the inputs and parameters that were used to create them.

For example, if we declare the predicate mtimes (declared as in § 2.2) to be tabled, and we have two matrix a and b, the first time mtimes(a,b,C) will be queried the Matlab engine will be called. Once the computation done, and the queried predicate has successfully been unified with mtimes(a,b,c), where c is actually a term representing the product of a and b, the corresponding tuple will be stored. When querying again mtimes(a,b,C), the computation will not be done, but the stored result will be returned instead. It also means that, given c, we can get back to a and b, using the query mtimes(A,B,c).

2.4 Implementation

Knowledge Machines are built on top of SWI Prolog[1] (which is one of the most *user-friendly* Prolog) and PostgreSQL[2]. The *workspace* they implement is accessible through a Prolog command-line, and new facts, composite formulæ or new evaluation engines can be directly asserted through it, or through external source files.

Each Knowledge Machine also wraps a component able to make it usable remotely. This can be seen as a simple Servlet, able to handle remote queries to local predicates, through simple HTTP GET requests. This will be useful when other components of the framework (such as the planner described in § 3.4) have a global view of the system and need to dynamically organise a set of Knowledge Machines.

3 A Semantic Web Knowledge Environment

In this section, we describe how we provide a shared and distributed knowledge environment, using Semantic Web technologies (see § 3.1) and a set of domain ontologies (see § 3.3). We will also explain how Knowledge Machines can interact with this environment in § 3.2, using entry doors generated using the tool described in § 3.4.

We will refer to several technologies, all part of the Semantic Web effort, which we will briefly overview here. RDF (Resource Description Framework[3]) defines how to describe resources (located by an Universal Resource Identifier[4]), and how to link them, using *triples* (sets of subject/predicate/object). For example, using RDF, I can express that the resource representing a given artist has produced several albums. An OWL (Ontology Web Language[5]) ontology is able to express knowledge about one particular domain by identifying its important concepts

---

[1] See http://www.swi-prolog.org/
[2] See http://www.postgresql.org/
[3] See http://www.w3.org/RDF/
[4] See http://www.gbiv.com/protocols/uri/rfc/rfc3986.html
[5] See http://www.w3.org/2004/OWL/

and relations, in RDF. SPARQL (Simple Protocol And RDF query language[6])
defines a way to query RDF data. Finally, a SPARQL *end-point* can be seen as
a public entry door to a set of RDF statements.

3.1   Why Use Semantic Web Technologies?

The 'metadata' Mistake. In this shared knowledge environment, we may
want to state circumstances surrounding the creation of a particular raw multimedia data. One option is to 'tag' each piece of primary data with further data,
commonly termed 'metadata', pertaining to its creation. For example, CDDB[7]
associates textual data with a CD, while ID3[8] tags allow information to be
attached to an MP3 file. The difficulty with this approach is the implicit hierarchy of data and metadata. The problem becomes acute if the metadata (eg the
artist) has its own 'meta-metadata' (such as a date of birth); if two songs are
by the same artist, a purely hierarchical data structure cannot ensure that the
'meta-metadata' for each instance of an artist agree. The obvious solution is to
keep a separate list of artists and their details, to which the song metadata now
refers. The further we go in this direction, i.e. creating new first-class entities for
people, songs, albums, record labels *etc.*, the more we approach a fully *relational*
data structure.

Towards a Scalable Solution. We also want this data structure to be *distributed.* Any entities contributing to the knowledge environment may want to
publish new assertions, eventually concerning objects defined in an other place.
This is why we are using RDF. We also want to be able to specify what types
of objects are going to be in the domain of discourse and what predicates are
going to be relevant. Designing an *ontology* [4] of a domain involves identifying
the important concepts and relations, and as such can help to bring some order
to the potentially chaotic collection of predicates that could be defined. We may
also want to dynamically introduce new domains in the knowledge environment.
This is why we are using OWL.

3.2   Integrating Knowledge Machines and the Knowledge
Environment

Querying the Semantic Web Within Knowledge Machines. Another
characteristic of the Knowledge Machines is the ability to query the Semantic
Web (through a set of *end-points*, as we will see in § 3.4) using SPARQL. Someone
working on a Knowledge Machine is able to create an *interpretation* of the *theory*
(OWL ontologies) in the form of a collection of predicates. It is then possible
to use these predicates when building composite formulæ in the language of
predicate calculus.

For example, we can imagine the following SPARQL query that associates an
audio file and the sampling rate of the corresponding digital signal:

---

[6] See http://www.w3.org/TR/rdf-sparql-query/
[7] See http://www.gracenote.com/
[8] See http://www.id3.org/

```
PREFIX mu:  <http://purl.org/NET/c4dm/music.owl#>
SELECT ?a ?r WHERE {
    ?a rdf:type mu:AudioFile. ?a mu:encodes ?dts.
    ?dts rdf:type mu:DigitalSignal.
    ?dts mu:samplingRate ?r }
```

We can associate to this query the following predicate, whose first argument will be bound to the audio file, and whose second argument will be bound to the corresponding sampling rate:

$$audiofile\_samplingrate(AudioFile, SamplingRate)$$

Now we can use this predicate in composite formulæ, perhaps to use this sampling rate information as one of the inputs of an algorithm wrapped in another predicate.

Exporting Knowledge to the Semantic Web. Someone working on a particular Knowledge Machine may want, at some point, to state that a particular predicate is relevant, according to the domain ontologies hold by the knowledge environment. This is equivalent to stating that this predicate has a particular *meaning* which can be expressed using one of the *vocabularies* we have access to. Thus, we want to be able to state a match between a predicate and a set of RDF triples. Moreover, we want to express this match either in the language of predicate calculus (in order to export new information when this predicate holds new knowledge) and in terms of OWL/RDF to allow automatic reasoning (as described in § 3.4) in the Semantic Web layer. We developed a simple ontology of *semantic matching* between a particular predicate and a conceptual graph. This ontology uses the RDF *reification*[9] mechanism in order to express things like '*this* predicate in *this* Knowledge Machine is able to create *these* RDF triples'. This can be seen as a limited subset of OWL-S[10], where the *effects* of a particular *process* can only consist in creating new RDF triples. For example, the predicate soxsr, able to change the sample rate of an audio file, can create some RDF triples, as represented in fig. 3. Thus, a Knowledge Machine can be represented as in fig. 4.

3.3 Domain Specific Ontologies

In order to make this knowledge environment understandable by all its actors (Knowledge Machines or any entities querying this environment), it needs to be designed according to a shared understanding of the specific domains we want to work on. An ontology can provide this common way of expressing statements in a particular domain. Such ontologies will be developed in the context of music in § 5.1. Moreover, the expressiveness of the different ontologies specifying this environment will implicitly state how dynamic the overall framework can be.

---

[9] See http://www.w3.org/TR/rdf-mt/#Reif
[10] http://www.daml.org/services/owl-s/

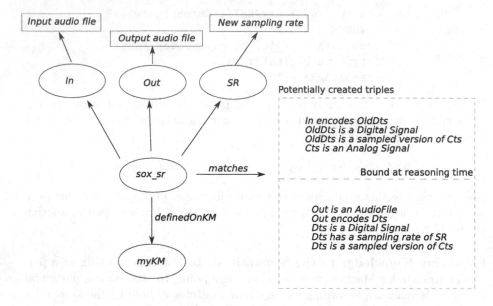

Fig. 3. Expressing a match between the predicate soxsr and the fact that it is able to change the sample rate of an audio file

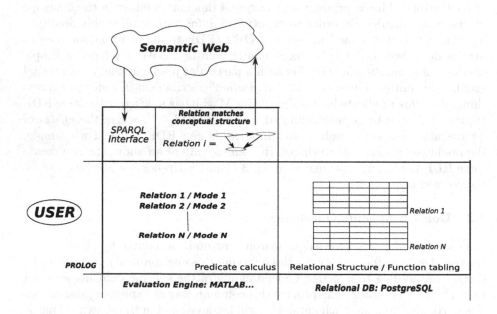

Fig. 4. Overall architecture of a Knowledge Machine

Indeed, the *semantic matching* ontology defined in the previous section has an expressiveness that directly depends on the different domain ontologies that are known.

For example, if we write an ontology that is expressive enough in the domain of football games and time sequences, and we have an algorithm which is able to segment a football match video (corner, penalty, ...), we will be able to express a conceptual match between *what* is done by the algorithm and a set of RDF statements conforming to this domain ontology.

However, in order to keep the overall framework in a consistent state, designing a new ontology must be done considering some points. These include modularity [5] and ontological 'hygiene' as addressed by the OntoClean methodology [6].

3.4 Handling Semantic Web Knowledge

At this point, we still need to make Semantic Web data available to both Knowledge Machines and other entities wanting to make queries.

XsbOWL: Creating SPARQL End-Points. In order to achieve this goal, we designed a program able to create SPARQL *end-points*: XsbOWL (see fig. 5). It allows SPARQL queries to be done through a simple HTTP GET request, on a set of RDF data. Moreover, new data can be added dynamically, using an other HTTP GET request.

Reasoning on Semantic Web Data. To handle reasoning on the underlying Semantic Web data, we bound XsbOWL to an XSB Prolog engine (which is more adapted to a large deductive database use case [7]). The latter, thanks to the inner XSB tabling mechanism, is able to provide reasoning on the positive entailment[11] subset of OWL Full[12]. XsbOWL is able to deal simultaneously with around 100000 RDF statements and still provides a really fast reasoning (less than 0.2 seconds per query). More scalability testing still has to be done.

Dynamically Exporting Knowledge to the Semantic Web. We also integrated a planner in this XSB engine, in order to fully use the information held by the *semantic matching* ontology. This one is planning which predicate it needs to call in which Knowledge Machine (using the remote calling mechanism described in § 2.4) in order to reach a *state of the world* (the set of all RDF statements known by the *end-point*) which will at least give one answer to the query (see fig. 6). For example, if there is a Knowledge Machine somewhere that defines a predicate able to locate all the segments corresponding to a penalty in a football match, querying the *end-point* for a sequence showing a penalty during a particular match should automatically use this predicate.

---

[11] See http://www.w3.org/TR/owl-test/
[12] See http://www.w3.org/TR/owl-ref/

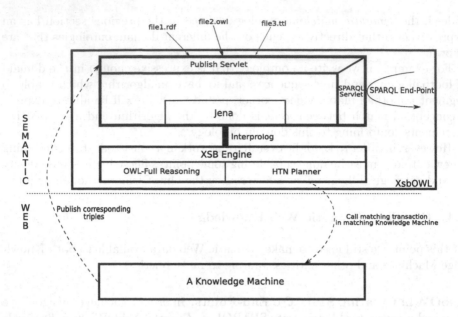

Fig. 5. XsbOWL: Able to create a SPARQL end-point for multimedia applications

4   Use Cases

In this section, we describe two different use cases, to give an insight of the wide range of possibilities the proposed knowledge management framework brings.

4.1   Enhanced Workspace for Multimedia Processing Researchers

Working in a Knowledge Machine environment to develop multimedia processing algorithm helps to create what we could call a *semantic workspace*. Every object is part of the same logical structure, based on predicate calculus. Moreover, the Knowledge Machine framework provides a brand new programming environment, aware of an *open context*. Therefore, while developing a new predicate, we may access knowledge perhaps already available or newly created by an other Knowledge Machine, and this in a completely transparent way.

While working on a multimedia feature extraction predicate, it is possible to access the knowledge environment *inside* the predicate. For example, while working on a melody extraction algorithm, we are able to state that a particular sub-algorithm is to be used if an audio signal was created by a particular instrument. This could lead to the transparent use of an instrument classification predicate exported by an other Knowledge Machine.

4.2   End-User Information Access

Once the shared information layer holds a substantial amount of knowledge, it can be useful for other entities (not part of the Knowledge Machines framework)

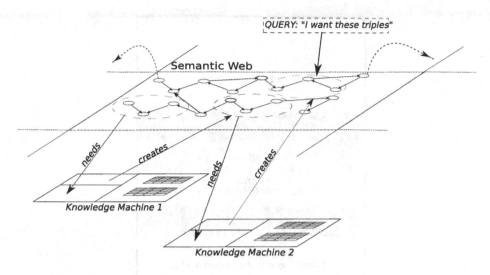

Fig. 6. Planning using the semantic matching ontology

to use a SPARQL end-point. For example, an interactive graphical viewer application (such as Sonic Visualiser[13]) should be able to submit simple queries to compute some features of interests (or to retrieve previously computed ones).

Moreover, as expressed in fig. 7, multimedia information retrieval applications can be built *on top* of this shared environment, through a layer interpreting the available knowledge. For example, if a Knowledge Machine is able to model the *textural* information of a musical audio file, and if there is an interpretation layer that is only able to compute an appropriate distance between two of these models, an application of similarity search can easily be built on top of all of this. We can also imagine more complex information access systems, where a large number of features computed by different Knowledge Machines can be combined with social networking data, all part of the shared information layer too.

5   Knowledge Management for Music Analysis

In this section, we will describe how this framework has been used for a music information management (this is explained in greater details in [2]). We will detail two Knowledge Machines, respectively dealing with format conversion and segmentation.

5.1   An Ontology of Music

Our ontology must cover a wide range of concepts, including non-physical entities such as a musical *opus*, human agents like composers and performers, physical

---

[13] See http://www.sonicvisualiser.org

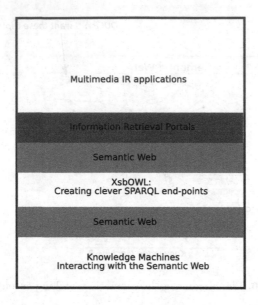

Fig. 7. The Multimedia Knowledge Management and Access Stack

events such as particular performances, informational objects like digital signals, and time. We will focus on the two main aspects of our ontology: physical events and time representation.

An Ontology of Events. Music production usually involves physical events that occur at a certain place and time and that can involve the participation of a number of physical objects both animate and inanimate.

The event representation we have adopted is based on the *token-reification* [8] approach. We consider an event occurence as a first class object or 'token', acting like a hook for additional information pertaining to the event. Regarding the on-tological status of event tokens, we consider them as being the way by which cog-nitive agents classify arbitrary regions of space-time. Our definition of an event is broad enough to include sounds (an acoustic field defined over some space-time region), performances, compositions, and even transduction and recording to produce a digital signal. We also consider the existence of *sub-events* to repre-sent information about complex events in a structured and non-ambiguous way. A complex event, perhaps involving many agents and instruments, can be broken into simpler sub-events, each of which can carry part of the information pertain-ing to the complex whole. For example, a group performance can be described in more detail by considering a number of parallel sub-events, each of which representing the participation of one performer using one musical instrument (see fig. 8 for some of the relevant classes and properties).

An Ontology of Time. Each event can be associated with a time-point or a time interval, which can either be given explicitly, *e.g.* 'the year 1963', or

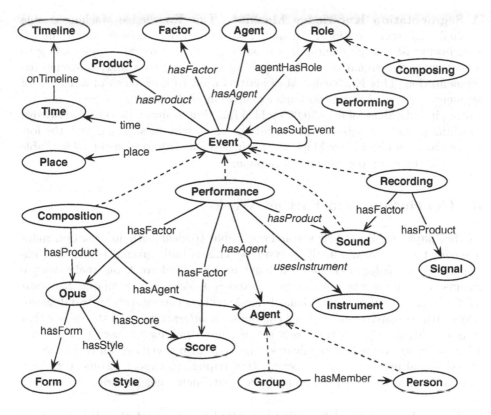

Fig. 8. Some of the top level classes in the music ontology

by specifying its temporal relationship with other intervals, *e.g.* 'during 1963'. Both must be related to a *timeline*, continuous or discrete, representing linear pieces of time which may be concrete—such as the one underlying a signal or an event, or more abstract—such as the one underlying a score. Two *timelines* may be related, using *timeline maps*. For example, an instance of this concept may represent the link between the continuous physical time of an audio signal and the discrete time of its digital representation.

5.2   Examples of Knowledge Machines

So far, two main Knowledge Machines are *exporting* knowledge to the shared knowledge environment.

A Format Conversion Knowledge Machine. The simplest one is about converting the format of raw audio data. Several predicates are exported, dealing with sample rate or bit rate conversion, and encoding. This is particularly useful, as it might be used to create, during the development of another Knowledge Machine, test sets in one particular format, or even to test the robustness of a particular algorithm to information loss.

A Segmentation Knowledge Machine. This Knowledge Machine is able to deal with segmentation from audio, as described in greater details in [9]. It exports just one predicate, able to split the time interval corresponding to a particular raw signal in several ones, corresponding to a machine-generated segmentation. This Knowledge Machine was used to keep track of hundreds of segmentations, enabling a thorough exploration of the parameter space, and resulting in a database of over 30,000 tabled function evaluations. This Knowledge Machine provides a segmentation ability on raw audio files. Along with the format conversion Knowledge Machine, it brings the ability to segment *all* available audio files by using our *planning* component.

6   Conclusions and Further Work

In this paper we described a framework able to deal with information management for multimedia analysis systems. This is built around two main components: *Knowledge Machines*, enabling to wrap and work on analysis algorithms using *predicate calculus* and *function tabling*, and a shared *Semantic Web* knowledge environment. Knowledge Machines can interact in two different ways with this environment. They can build an *interpretation* of the theory that it is holding at a given time as a set of logical predicates, thus to use them when building compound predicates. They can also specify a *match* between a set of logical predicates and a set of RDF triples, to export results to the Semantic Web layer, or to dynamically compute new ones in order to satisfy a query.

We can think of the Knowledge Machine framework as an artificial way of accessing *concepts*, which are defined by the domain ontologies held by the shared knowledge environment. Thus, a network of Knowledge Machines can bring an artificial and approximate *cognition* for multimedia related materials, against a *culture* which is defined by the different ontologies. It leads to a distributed intelligence, as mentioned in [10], and thus is similar in some ways to agent technologies. However, it brings an artificial *cognition* of available multimedia materials instead of *services*—getting things done.

There are several possible extensions to this framework, such as handling *trust* in the Semantic Web environment. For example, we may want to express that a computer-generated segmentation of an audio file is less accurate than a human-generated one, and we may also want to *quantify* this accuracy. We could also do a statistical analysis to judge whether or not a particular algorithm has successfully captured a given concept, and if so, to declare a match between a wrapping predicate and this concept so that the algorithm gains a semantic value; subsequent queries involving this concept would then be able to invoke that algorithm (using the planner component) even if no key annotations are present in the shared environment. This would be an example of 'closing the semantic gap'.

Acknowledgments

The authors acknowledge the support of both the Centre For Digital Music and the Department of Computer Science at Queen Mary University of London for the studentship for Yves Raimond.

References

1. D. S. Weld, "Recent advances in ai planning," *AI Magazine*, 1999.
2. S. Abdallah, Y. Raimond, and M. Sandler, "An ontology-based approach to information management for music analysis systems," in *Proceedings of 120th AES convention*, 2006.
3. E. F. Codd, "A relational model of data for large shared data banks," *Communications of the ACM*, vol. 13, no. 6, pp. 377–387, 1970.
4. F. Baader, I. Horrocks, and U. Sattler, "Description logics as ontology languages for the semantic web," in *Essays in Honor of Jörg Siekmann*, ser. Lecture Notes in Artificial Intelligence, D. Hutter and W. Stephan, Eds.   Springer, 2003.
5. A. L. Rector, "Modularisation of domain ontologies implemented in description logics and related formalisms including owl," in *Proceedings of the international conference on Knowledge capture.*   ACM Press, 2003, pp. 121–128.
6. N. Guarino and C. Welty, "Evaluating ontological decisions with ONTOCLEAN," *Communications of the ACM*, vol. 45, no. 2, pp. 61–65, 2002.
7. K. Sagonas, T. Swift, and D. S. Warren, "XSB as an efficient deductive database engine," in *SIGMOD '94: Proceedings of the 1994 ACM SIGMOD international conference on Management of data.*   New York, NY, USA: ACM Press, 1994, pp. 442–453.
8. M. P. Shanahan, "The event calculus explained," in *Artificial Intelligence Today, Lecture Notes in AI no. 1600*, M. J. Woolridge and M. Veloso, Eds.   Springer, 1999, pp. 409–430.
9. S. Abdallah, K. Noland, M. Sandler, M. Casey, and C. Rhodes, "Theory and evaluation of a bayesian music structure extractor," in *Proceedings of the Sixth International Conference on Music Information Retrieval*, J. D. Reiss and G. A. Wiggins, Eds., 2005, pp. 420–425.
10. J. Bryson, D. Martin, S. McIlraith, and L. Stein, "Toward behavioral intelligence in the semantic web," *IEEE Computer, Special Issue on Web Intelligence*, vol. 35, no. 11, pp. 48–55, November 2002.

In Search for Your Own Virtual Individual

L. Moccozet[1], A. Garcia-Rojas[2], F. Vexo[2], D. Thalmann[2],
and N. Magnenat-Thalmann[1]

[1] MIRALab, University of Geneva, Route de Drize, 7, Site de Battelle
CH1227 Carouge/Genève, Switzerland
{moccozet, thalmann}@miralab.unige.ch
[2] VRLab, EPFL, Lausanne
CH-1215 Lausanne, Switzerland
{alejandra.garciarojas, frederic.vexo, daniel.thalmann}@epfl.ch

Abstract. The use of inhabited Virtual Environments is continuously growing. People can embody a human-like avatar to participate inside these Virtual Environments or they can have personalized character acting as mediator; sometimes they can even customize it to some extent. Those Virtual Characters belong to the software owner, but they could be potentially shared, exchanged and individualized between participants, such as already proposed by Sony with Station Exchange. Technology with standards could significantly improve the exchange, the reuse and the creation of such Virtual Characters. However an optimal reuse is only possible if the main components of the characters: geometry, morphology, animation and behavior, are annotated with semantics. This may allow to users searching for specific models and customize them. Moreover search technology based on the Web Ontology Language (OWL) can be implemented to provide this type of service. In this paper we present the considerations to build an ontology that fulfills the mentioned proposes.

Keywords: 3D shapes, Virtual Environments, Virtual Human, Standards, Ontology, Web Ontology Language.

1 Introduction

Complex 3D shapes such as Virtual Humans are becoming more and more popular in various areas. One of them is the game industry, where players are represented with human like avatars in immersed in a Virtual Environments.

On one hand, with the successful development of massively multiplayer online role playing games or MMORPG and their persistent virtual worlds, it becomes common to exchange or to buy 3D objects such as virtual weapons and tools or even 3D virtual human characters used as avatars. For example, Sony is proposing Station Exchange [1] (Fig. 1), to allow players exchanging virtual characters. On the other hand, the recent advent of efficient and accurate acquisition systems and reconstruction platforms has potentially increased the availability and dissemination of complex digital shapes such as Virtual Humans [2, 3]. As a consequence, the supply and demand for Virtual Human will probably rapidly expand. However, the

Y. Avrithis et al. (Eds.): SAMT 2006, LNCS 4306, pp. 26–40, 2006.
© Springer-Verlag Berlin Heidelberg 2006

infrastructures that would be required for sharing large databases of Virtual Human models and populating Virtual Environments are still not mature; moreover a common understanding to share this models and their information does not exist.

Fig. 1. Snapshot from a Virtual Character proposed for auction on the Sony Station Exchange web server

Considering a Virtual Environment, the individualization of character is possible, but it is limited to some body shapes and changes of skin and hears colors, garments, accessories, etc. a large variety is not accessible. This is because it is expensive for designers to create a large number of characters and accessories. As a consequence, we may consider other techniques for the creation of virtual humans such as scanning and example-based or knowledge-based methods [2, 3] (Fig. 2). Therefore we have more options for shape creation.

The conventional shape acquisition and processing pipeline severely neglects to manage and take advantage of the knowledge associated to shapes as it would be required to efficiently share and interoperate their digital representation. Moreover, digital shapes such as Virtual Humans include different layers of semantic information. A Virtual Human is usually considered to a limited extent as a 3D shape with control animation structures. This representation only counts for a physical representation and movement control of the Virtual Human. However Virtual Human models should not only provide a physical representation of Humans: they need to personify Humans as described by Kshirsagar et al. in [4]. Processing a 3D model of a human to its virtual similar can be seen as anthropomorphism or personification. Anthropomorphism[1], also referred to as personification[2] or prosopopoeia, means to concede human attributes to inanimate objects. These attributes may include sensations, emotions, desires, physical gestures and expressions, and powers of speech, among others. In our context it consists in representing a virtual counterpart of a real or "potential" Human from a virtual 3D shape.

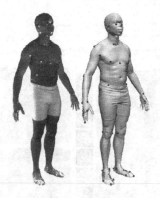

Fig. 2. Example of a scanned body shape from the CAESAR database

Such a representation ideally requires embedding different levels of knowledge:

- Geometric: What is the mesh resolution of the Virtual Human shape?
- Animation: What is the level of articulation of the Virtual Human model and is it able to perform a specific given sequence of motion?
- Morphological: What is the size, gender, or weight of a Virtual Human model?
- Behavioral: What are the skills, the mental state or the mood of a Virtual Human model?

These levels are intricately related: the size of a Virtual Human can be computed from the 3D shape of the body; the mental state will influence the gestures and motion that the Virtual Human can perform and the way it will perform them.

Current 3D search engines such as the Princeton 3D models search engine [5-8], are based on a few geometric metadata and 2D/3D shapes similarity and they are not able to process Virtual Human models deeper than at the geometric level [9]. There exists an obvious need to represent and store a Virtual Human model as a virtual personification of Humans. This representation should include the different layers of knowledge and semantic information so that every Virtual Human model can be efficiently searched and shared for populating Virtual Environments with avatars or autonomous characters.

---

[1] http://en.wikipedia.org/wiki/Anthropomorphic
[2] http://en.wikipedia.org/wiki/Personification

Search is commonly performed using keywords or specific values of fields in the databases. This way of searching is very limited for complex multimedia objects such as Virtual Humans. For improving it we propose to use Ontologies, the technology used in the semantic web. The most complete language to express semantics is Ontology Web Language (OWL) [10]. This language is created upon Resource Definition Framework (RDF) [11]. OWL is a more expressive language than RFD, it provides relations between concepts in a logic way. Therefore it is possible, if implemented in a search engine, to make queries in natural language with a detailed criterion. In the sections 2 and 3 of this paper, we present the principal considerations for creating virtual individuals and some existing standards and good practice within these representations. In section 4 we present the structure of the proposed ontology. Using this ontology, we present a usage scenario that creates an individual from a semantic description is presented in section 5. Finally the conclusions and future work are presented in section 6.

2 Virtual Human Personification

The personification process [12] in virtual humans includes mainly 3 stages:

2.1 Physical Personification

This aspect is related to the visual appearance of the Virtual Humans. We can consider 2 levels. The primary one that is mandatory regarding the human body itself and the secondary one that is optional regarding the

2.1.1 Primary Physical Personification
This personification step is dedicated to the modeling of 3D shape of the body. Although any 3D modeling software would provide the basic functionalities to create a 3D human body shape, new methods have been recently investigated in order to provide more intuitive approaches based on: parameterized human models; fitting predefined template or existing knowledge [13-17]. The scanning technology allows improving these new approaches by providing large amount of real data [2, 3]. Template-based method can be therefore improved by using statistical analysis methods to optimize the parameterization and control of the template model.

2.1.2 Secondary Physical Personification
This step is dedicated to an optional personification step, which consists in attaching accessories and artifacts (explicit hair, clothes, tools, jewels...) to the body shape. Accessories and artifacts are 3D shapes that are attached to some specific location of the body shape using landmarks.

Both categories refer to 3D geometry and share common characteristics. These characteristics refer those of any 3D shape; they can be: number of vertex, number of edges, a classification (mesh, contour, curve, etc), scale factor, material, texture, etc. Accessories and artifacts are connected to the body shape using landmarks that are located on the body surface. These landmarks are mainly 3D points or eventually 3D curves.

2.2 Expressional Personification

Here we consider all aspects to give movement to a shape. As mentioned before for animating a character there is an implementation of articulation (skeletal structure) that drives the shape. As a result we have animation values to give to the articulations and shape deformation (skin binding). Shape deformation may be attached to the shape, or inside an algorithm. To produce animation values exist many techniques, we can split them in motion capture and key frame [18]:

2.2.1 Motion Capture
This animation technique is made with recorded movements of a real person using sensors in his body. This technique is more natural, but it is not easy to modify it; normally it is attached to a specific anatomy, and a lot retargeting work has to be made for reusing it.

2.2.2 Keyframe Animation
It is mainly defined as important frames that record the values of articulations, and the computer calculates with an interpolator the frames in between to produce the animation. This kind of animation can use other techniques as Inverse kinematics or dynamics. Its advantage is that one has high control of the animation but they are not realistic.

2.3 Logical and Emotional Personification

This is an open issue in research, where the results are the implementation of different kind of models to simulate behaviors [19, 20]. This model depends on the level of autonomy of the character. While less level of autonomy more complex is the model. Most of the created models are for completely autonomous characters. However those models have some characteristics in common, like the use of psychological models to represent affect (personality, emotion, mood, etc). It is important the representation of this individual characteristics, because they can be used to give expressions in animation, like happiness, anger, tiredness, etc. In fact some emotions are identified to drive synthetic animations [21].

We have mentioned the process that is followed in the creation of individual virtual humans. In the next section we will translate this process in terms of computational representation in a common practice.

3 Virtual Human Representations

Existing representations for Virtual Humans can be classified into two main categories: physical personification representations and anthropomorphic representations. The most developed is dedicated to physical representation and it is principally intended to describe the body shape, and control with joint motions or key frame animations. The other category is dedicated to provide higher level of personification with human-like attributes such as anthropometric or personality descriptors. Many of them are oriented to offer some kind of scripting animation language that allows defining expressive animations (i.e. animations that depicts emotions and behavior). None of these representations offers a common ground to integrate and cover all the aspects related to the personification of a Virtual Human models.

3.1 Physical Personification Representation

Most of these models are focused towards the representation and storage of:

- One or more 3D shapes that represent the outer shape of the body,
- Control animation structures, which include mainly an articulated control skeleton for the body.
- Skin binding information, which defines the way skin meshes are controlled and deformed according to skeleton motions.

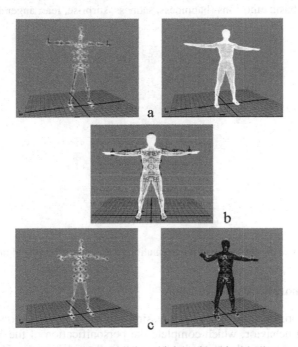

Fig. 3. (a) Skeleton (left) and skin modeling (right), (b) skin binding to the skeleton, (c) skeleton motion (left) and skin deformation (right)

These models are limited to the representation of the physical and expressional personification levels. They provide a low level of personification and ensure that the Virtual Human model look and move similarly to a real one. The main issue is to model an articulated skeleton that is able to reproduce as close as possible the degrees of freedom of the anatomical skeleton and to keep the 3D skin realistically adjusted around the skeleton for any posture with Skeleton Driven Deformation [22, 23] (Fig. 3).

The best way to share this physical representation is through standards. In computer graphics exists the standardization for the animation structure in H-Anim [24] standard. This structure mainly represents a simplified human skeleton with some specifications of the name, number and components of human articulations, as presented in the figure 4. This standard has been used or could be used in many 3D exchange formats like MPEG-4 [25], VRML/X3D [26], Collada [27], CAL3D [28], BVH [29], etc. It provides an articulated structure with standard nomenclature and best practice. A typical example of the benefit of such standards is that an animation sequence can be shared among different 3D characters if they use H-Anim standard as animation control structure.

Still these formats can address much more detailed information of Virtual Human, like the level of articulation, level of detail, joint limit, joint coupling, etc. This kind of information is impossible to know at first instance, and they are important at the time of animating. MPEG-4 is also considering the face and facial animation and provides a scheme to control facial expressions based on control points. Facial animation in MPEG-4 also provides premises for higher level representation by taking into account six basic emotions (happiness, sadness, surprise, fear, anger and disgust).

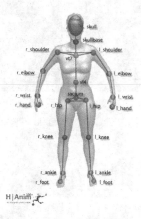

Fig. 4. H-ANIM standard representation or articulated Virtual Human

3.2 Anthropomorphic Representations

This type of representation allows addressing higher levels of knowledge, mainly morphology and behavior, which complete the personification of the Virtual Human models. A typical example of such representation model is the Human Markup Language. These representations are either too much targeted for a specific type of applications or too much abstract regarding the physical representation aspect.

3.2.1 Multi-purposes Representations
The Human Markup Language is an XML based representation proposed by the OASIS Human Markup Technical Committee. According to the Human Markup Language Primary Base Specification 1.0 [30, 31], the scope of the Human Markup Language is quite broad as it targets applications such as enabling real-time animated behaviors for 3D representations of humans to enhancing diplomatic communication with translation services and/or applications capable of making provisions for cultural practices. The Human Markup Language is expected to describe a fundamental set of

characteristics of human entities and human activities as they occur in digital information systems. All these characteristics can be applied to represent artificial humans as well as real humans.

The Human ML basically includes simple types, such as age, gender or physical descriptors (weight, hair color, eye color...) and complex types such as Address, Human artifact (clothes, jewels...), Belief, Human Communication Channel, Community as an Abstract Human Organization, Human Culture, Human Emotion, Geolocator, Haptic (defined as the strength, location, and body part used in a touching behavior), Human Intent, Kinesic (Human Movements), Human Personality Type...

3.2.2 Anthropometric Representations
For representations such as the one proposed in [32], the target is CAD-CAM simulations involving human activities. The objective is to be able to query an inhabited Virtual Environment for controlling and analyzing CAD-CAM simulations. The proposal for representing Virtual Humans is focused towards the anthropometric attributes. It considers 1) human information: gender, stature, age, mass, joints, clothing, nationality...; 2) human's state: orientation, location, direction of motion, posture, joint angle... It also takes into account clothing, objects and environment.

3.2.3 Behavioral Representations
Other specific representations have been proposed with a more focused scope of application. Most of them are more scripting languages dedicated to represent animation of Virtual Human than representing the Virtual Human itself. The Character Mark-up Language (CML) is an XML-based scripting language proposed to link available engines for generating and controlling believable behavior of animated agents with the corresponding animated representations [33]. The Avatar Markup Language (AML), also based on XML, aims at encapsulating the Text to Speech, Facial Animation and Body Animation in a unified manner with appropriate synchronization [34]. The Emotion Annotation and Representation Language (EARL) [35] is focused in the behavioral aspects Emotion-oriented (or "affective") computing. They provide an annotation of emotions like intensity, variation, confidence, etc.; with the goal of interpretation and generation of behaviors.

Those XML languages offer a description for the representation of the concepts; they are adequate for annotation, and to provide a standardization of terms. According to the semantic stack [36] the next step is to define an ontology to provide a semantic meaning of the content.

4 Ontological Representation of Virtual Humans

Once we have a data structure and data concepts we can specify the relation of concepts to build the formalization of the knowledge. A diagram of the ontology is presented in the Figure 5. The concepts that are circled are subclasses of the concept resource (in the bottom of the diagram). Resource is the content that the user gets when searching for something. Resources have properties like file info (extension, size, URL), version, Author, creation date, etc. Some of the resources can be found in separated files or more than one in the same file. For the case of behavior controller the user can get an algorithm.

Each shape representation we identified in the previous sections is represented as a concept; they are described in the ontology as follows:

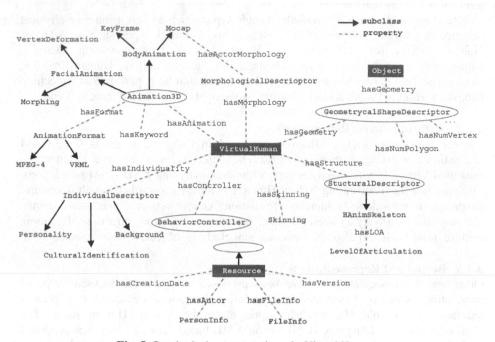

Fig. 5. Ontological representation of a Virtual Human

Geometry (hasGeometry, hasObject, hasGarment): The geometry is the physical visual representation of the Virtual Human and his two aspects, primary: body shape, and secondary: accessories, garments, etc. Therefore, we can create a general description for any kind of geometry shape and heritage its properties to specialized subclasses, specifically for Virtual Humans or objects. This common class contains the geometric properties: number of vertex, edges, scale, material and texture. The Virtual Human has a Geometrical Representation where we can know the kind of geometry it has (e.g. Surface mesh).

Animation (hasStructure, hasAnimation): The animation of Virtual Human can be for face or body. Each one has a format defined (MPEG-4, VRML, etc.), The way of animating each other is different, and it is possible to specialize the content of a face animation; for example if we have a happiness expression we can classify the animation values [37]. Body animation can be classified in KeyFrame or MoCap (Fig. 6). In order that an animation can take place, the character model should share the same kind of structure as the animation. This structure is defined in the Structural Descriptor (Fig. 7). We have based this structure in the H-Anim standard, which in fact is adopted in most of the geometry and animation file formats; however, any

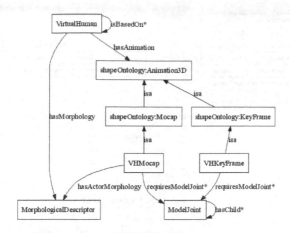

Fig. 6. Overview of the animation descriptor in the ontology

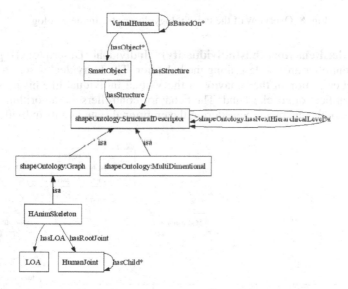

Fig. 7. Overview of the animation descriptor in the ontology

other structures can be also represented. The Motion Capture animation has also some special properties such as the morphology of the actor. This can be useful to predict and support the work one has to do in order to retarget an animation for a model with a different morphology.

Morphology (hasMorphology): Morphological Descriptor (Fig. 8) contains the information like: Age, weight, height, gender. This information is used to describe the Virtual Human body shape as people are used to do it for real human beings.

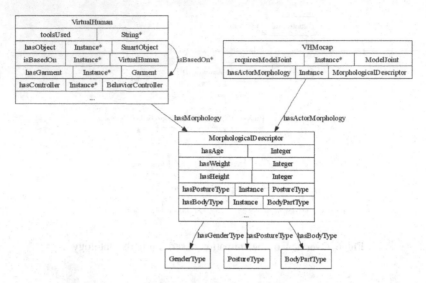

Fig. 8. Overview of the morphology descriptor in the ontology

Behavior (hasBehavior, hasIndividuality): Individual Descriptor (Fig. 9) and Behavior controller are for describing the behavior. The Individual descriptor contains the constant definition of the behavior of the Virtual Individual like his personality or cultural identification, background. The behavior controllers are algorithms that drive the behavior of the character considering the emotional state and its individuality.

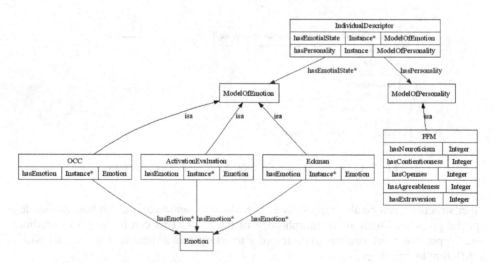

Fig. 9. Overview of the individuality descriptor in the ontology

This is a simplified version of the ontology with the most important definitions we have explained in this paper; however much more information has been considered, for example the structure of H-Anim (joints and segments), the animation formats of

MPEG-4 (FAP, BAP), etc. Finally we can see in the diagram that there is a lot of information around one concept, the Virtual Human. There are physical concepts and abstract concepts that give the meaning to the Virtual Human. In the next section we will present a case when a user wants to build its own virtual character from scratch by searching the desired components in the ontology.

5 Create a Virtual Individual

To give an example of the creation of a virtual individual we present the process we may follow for playing a specific type of motion animation sequence. Here we have used Protégé [38] to create the instances of some of our resources. Also in Protégé with the nRQL [39] plug-in we search for the components we want.

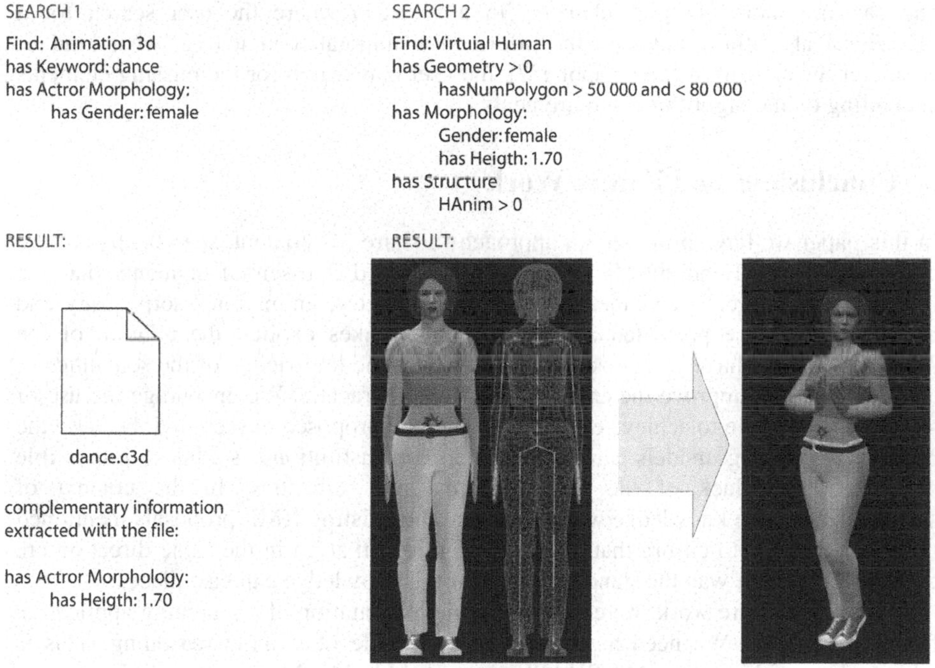

SEARCH 1

Find: Animation 3d
has Keyword: dance
has Actror Morphology:
    has Gender: female

SEARCH 2

Find: Virtuial Human
has Geometry > 0
    hasNumPolygon > 50 000 and < 80 000
has Morphology:
    Gender: female
    has Heigth: 1.70
has Structure
    HAnim > 0

RESULT:

dance.c3d

complementary information
extracted with the file:

has Actror Morphology:
    has Heigth: 1.70

RESULT:

Fig. 10. Example of the creation of virtual individual

This example is illustrated in the figure 10. Let's say that our designer, Mireille, wants an animation of a woman dancing; she looks in the ontology for animations of women that have the keyword dancing. She found one in C3D [40] format which can be open it in Motion Builder [41];, there she makes a retargeting of the MoCap animation to the H-anim skeleton, and she exports the animation to 3DMax [42]. There, she can export the animation in our proprietary format WRK, using a home made plug-in, this format uses H-Anim standard. She found that the mocap animation is made for a woman of 1.70 m. Therefore, she needs to search for a woman with a height of 1.70 m, with a structure of H-Anim, and also with a number of polygons between 50k and 80k, because that is the range we can load models and animate in our viewer without problems. She found a model in Collada format that can be opened in 3DMax, and exported in the format we use in the lab. Finally we can load both model in a 3D sequence viewer and play the animation.

This process can take 30 minutes in the best case, and much more depending on how "clean" are files that the designer got. We have noticed the benefit it could imply if we simplify the work using an optimized search like the one above. Currently a designer would first need to locate and identify the dancing sequence by playing each of the available animation sequences, and then to check if the dancing sequence fits the selected human model. They may have to do a lot of retargeting job because the scale of the model it was made the animation is not the same as the model we want to apply to, and much more other issues.

More complicated search can take more time in the adaptation. For example, when one can use more complex elements in a scenario where the user search for a behavioral algorithm, and with the information annotated in it (e.g. what kind of character the algorithm can be applied?), the user can search for the missing elements depending on the algorithm's requirements.

6 Conclusions and Future Work

In this paper we have proposed an approach to share 3D content, specifically for the creation of virtual individuals. We have considered 3 essential elements that are involved in the creation of these characters: geometry, animation, morphology and behavior. We have presented an ontology that makes explicit the relation of the elements, and we have presented a scenario where the knowledge of the semantics of the elements can improve the creation of virtual characters. We encourage the use of standards to be able to achieve examples as the one proposed in section 5, because the process of sharing models among designers or institutions is almost impossible because of the lack of the use standards and semantics. In the context of standardization the knowledge we may look at the existing XML proposals mentioned in the section 3.2 to ensure that other contributions that go in the same direction are compatible. In that way the standardization of the knowledge can take place.

Within our future work, we consider the implementation of a searching application for this ontology. We need a search engine capable of making reasoning. This is taking place within the AIM@SHAPE network [43, 44]. Moreover, a difficult issue that needs to be faced is at the moment of extraction of metadata from resources to populate the ontology. However, the extraction of the metadata form the most of the 3D files can be made with applications to automate the process.

Acknowledgements

This research has been funded by the EC and the Swiss Federal Office for Education and Science in the framework of the European Networks of Excellence IST-AIM@SHAPE (http://www.aim-atshape.net) and IST-HUMAINE (http://emotionresearch.net).

References

1. *Station Exchange.* http://stationexchange.station.sony.com/.
2. B. Allen, B. Curless, and Z. Popovic. *The space of human body shapes: reconstruction and parameterization from range scans.* in *SIGGRAPH '03.* 2003.
3. H. Seo and N. Magnenat-Thalmann, *An Example-Based Approach to Human Body Manipulation*, in *Graphical Models*, A. Press, Editor. 2004. p. 1-23.
4. S. Kshirsagar and N. Magnenat-Thalmann. *A Multilayer Personality Model.* in *2nd International Symposium on Smart Graphics.* 2002.
5. Princeton3D, *models search engine.* http://shape.cs.princeton.edu/search.html.
6. O. IV. http://ship.nime.ac.jp/~motofumi/Ogden/.
7. D. Search. http://3d-search.iti.gr/3DSearch.
8. C.s. Characteristics, *Content-based Classification of 3D-models.* http://merkur01.inf.uni-konstanz.de/CCCC/.
9. R.C. Veltkamp and J.W.H. Tangelder, eds. *Content Based 3D Shape Retrieval.* Encyclopedia of Multimedia. 2006, Springer Borko Furht (Ed.).
10. OWL, *Web Ontology Language Overview.* http://www.w3.org/TR/owl-features/.
11. RDF, *Resource Description Framework.* http://www.w3.org/RDF/.
12. N. Magnenat-Thalmann and D. Thalmann, *Handbook of Virtual Humans*, ed. Wiley. 2004.
13. C. C. and L. Wang, *Parameterization and parametric design of mannequins*, in *Computer-Aided Design.* 2005. p. 83-98.
14. W. Lee, J. Gu, and N. Magnenat-Thalmann, *Generating animatable three-dimensional virtual humans from photographs.* Computer Graphic Forum 2000. **19**(3): p. 1–10.
15. K.-M. Tong, K.-C. Hui, C. C., and L. Wang. *Mesh Fitting Based 3D Character Modeling.* in *Edutainment 2006.* 2006.
16. P.P.J. Sloan, C.F. Rose, and M.F. Cohen. *Shape by example.* in *Symposium on Interactive 3D Graphics.* 2001.
17. MAKEHUMAN. http://www.dedalo3d.com/index.php?filename=SXCOL/makehuman/abstract.html.
18. R. Parent, *Computer Animation: Algorithms and Techniques*, ed. Morgan-Kaufmann. 2001, San Francisco.
19. E.d. Sevin and D. Thalmann. *An Affective Model of Action Selection for Virtual Humans.* in *Agents that Want and Like: Motivational and Emotional Roots of Cognition and Action symposium at the Artificial Intelligence and Social Behaviors.* 2005. Hatfield, England.
20. T. Conde and D. Thalmann. *Autonomous Virtual Agents Learning a Cognitive Model and Evolving.* in *Intelligent Virtual Agents, Lecture Notes in Computer Science.* 2005: Springer-Verlag
21. P. Eckman, *Facial expression and emotion*, in *Am Psychologist.* 1993. p. 384–392.
22. J.P. Lewis, M. Cordner, and N. Fong. *Pose Space Deformations: A Unified Approach to Shape Interpolation and Skeleton-Driven Deformation.* in *SIGGRAPH '00* 2000: Addison-Wesley.
23. B. Allen, B. Curless, and Z. Popovic. *Articulated body deformation from range scan data.* in *SIGGRAPH 2002.* 2002.
24. H-anim., *The humanoid animation working group.* http://www.h-anim.org.
25. MPEG-4, *Overview of the MPEG-4 standard.* http://www.chiariglione.org/mpeg/standards/mpeg-4/mpeg-4.htm.
26. *Web 3D consortium.* http://www.web3d.org/.
27. COLLADA, *Asset Exchange Schema for Interactive 3D.* http://www.khronos.org/collada/.
28. CAL3D. http://home.gna.org/cal3d/.

29. Biovision, *BVH*. http://www.cs.wisc.edu/graphics/Courses/cs-838-1999/Jeff/BVH.html.
30. J. Landrum and S.C.d. Ram. *Human Markup Language [HumanML]: Humanness Content and Sharing across Perspective Shifts*. in *CIDOC CRM Special Interest GroupConference*. 2003. Washington, D.C.
31. J. Peltz, *HumanML: The Vision, Jay Peltz*, in *DM Direct Newsletter*. 2005 http://www.dmreview.com/article_sub.cfm?articleID=1033534.
32. J.D. Ianni. *Standardizing Human Model Queries*. in *Digital Human Modeling For Design And Engineering Conference And Exhibition*. 2001. Arlington, VA, USA.
33. Y. Arafa and A. Mamdani. *Scripting Embodied Agents Behavior with CML: Character Markup Language*. in *International Conference on Intelligent User Interfaces*. 2003. New York: ACM Press.
34. S. Kshirsagar, A.Guye-Vuilleme, K. Kamyab, N. Magnenat-Thalmann, D. Thalmann, and E. Mamdani. *Avatar Markup Language*. in *8th Eurographics Workshop on Virtual Environments*. 2002.
35. M. Schröder, H. Pirker, and M. Lamolle. *First suggestions for an Emotion Annotation and Representation Language*. in *Language Resources and Evaluation*. 2006. Genoa, Italy.
36. W3C, *Semantic stack architecture*. http://www.w3.org/2000/Talks/1206-xml2k-tbl/slide10-0.html.
37. F. Vexo A. Garcia-Rojas, D. Thalmann, A. Raouzaiou, K.Karpouzis, S. Kollias, L. Moccozet and N. Magnenat-Thalmann., *Emotional Face Expression Profiles Supported by Virtual Human Ontology*. Computer Animation and Virtual Worlds Journal, 2006. **17**(Special Issue): p. 259-269.
38. Protégé, *Stanford medical informatics*. 2005 http://protege.stanford.edu/index.html.
39. nRQL, *New racer query language (nrql) interface to owl ntologies in Protégé*. http://www.cs.concordia.ca/~k_bhoopa/nrql.html.
40. C3D, *Coordinate 3D*. http://www.c3d.org/.
41. Ronan Boulic, Ramon Mas, and Daniel Thalmann, *A robust approach for the control of the center of mass with inverse kinetics*. Computers & Graphics, 1996. **20**(5): p. 693-701.
42. Autodesk 3ds Max. http://usa.autodesk.com/adsk/servlet/index?id=5659302&siteID=123112.
43. R. Albertoni, L. Papaleo, M. Pitikakis, F. Robbiano, M. Spagnuolo, and G. Vasilakis, *Ontology-based Searching Framework for Digital Shapes*, in *Applications of Semantic Web II (SWWS)* L.N.i.C. Science, Editor. 2005, Springer-Verlag GmbH. p. 896-905.
44. George Vasilakis, Marios Pitikakis, Manolis Vavalis, and Catherine Houstis. *A Semantic Based Search Engine for 3D Shapes: Design and Early Prototype Implementation*. in *2nd European Workshop on the Integration of Knowledge, Semantic and Digital Media Technologies (EWIMT2005)*. 2005. http://www.iee.org/oncomms/pn/visualinformation/ewimt2005_0067.pdf.

Enabling Multimedia Metadata Interoperability by Defining Formal Semantics of MPEG-7 Profiles

Raphaël Troncy[1],
Werner Bailer[2], Michael Hausenblas[2], Philip Hofmair[2], and Rudolf Schlatte[2]

[1] CWI Amsterdam, P.O. Box 94079, 1090 GB Amsterdam, The Netherlands
`raphael.troncy@cwi.nl`
[2] JOANNEUM RESEARCH Forschungsgesellschaft mbH, Institute of Information Systems and Information Management, Steyrergasse 17, 8010 Graz, Austria
`firstName.lastName@joanneum.at`

Abstract. MPEG-7 can be used to create complex and comprehensive metadata descriptions of multimedia content. Since MPEG-7 is defined in terms of an XML schema, the semantics of its elements have no formal grounding. In addition, certain features can be described in multiple ways. MPEG-7 profiles are subsets of the standard that apply to specific application areas, which can be used to reduce this syntactic variability, but they still lack formal semantics. In this paper, we propose an approach for expressing semantics explicitly by formalizing the semantic constraints of a profile using ontologies and rules, thus enabling interoperability and automatic use for MPEG-7 based applications. We demonstrate the feasibility of the approach by implementing a validation service for a subset of the semantic constraints of the Detailed Audiovisual Profile (DAVP).

1 Introduction

The amount of multimedia data being produced, processed and consumed is growing, as is the number of applications dealing with multimedia content. In many of these applications, metadata descriptions of the content are important. MPEG-7 [13], formally named Multimedia Content Description Interface, is designed as a standard for representing such descriptions in a broad range of applications. In order to cover diverse requirements scenarios [16], many *descriptors* and *descriptions schemes*, as well as the relationships between them, have been defined. The descriptors and description schemes are together referred to as *description tools*, and a description is a particular instantiation of these. There are description tools for diverse types of annotations on different semantic levels, ranging from very low-level features, such as visual (e.g. texture, camera motion) or audio (e.g. melody), to more abstract descriptions. The flexibility of MPEG-7 is very much based on structuring tools, which allow descriptions to be associated with arbitrary multimedia segments or regions, at any level of granularity, using different levels of abstraction.

Y. Avrithis et al. (Eds.): SAMT 2006, LNCS 4306, pp. 41–55, 2006.
© Springer-Verlag Berlin Heidelberg 2006

The downside of the breadth targeted by MPEG-7 is its complexity and its fuzziness [1,19,21]. For example, very different syntactic variations may be used in multimedia descriptions with the same semantics, while remaining valid MPEG-7 descriptions. Given that the standard does not provide a formal semantics for these descriptions, this syntax variability causes serious interoperability issues for multimedia processing and exchange, for example on the Web. To reduce this syntax variability, MPEG-7 has introduced the notion of *profiles* that constrain the way the multimedia descriptions should be represented for particular applications. However, these additional constraints, such as the MPEG-7 description tools, can only be represented in XML Schema, which allows only very limited control over the semantics of the descriptions [9,14,18]. Because of this lack of formal semantics, the resulting interoperability problems prevent an effective use of MPEG-7 as a language for describing multimedia.

In this paper, we propose a method to formalize the semantic constraints of an MPEG-7 profile and a semantic validation service using the formalization. In contrast to other work [4,9,18,20], we do not intend to completely map the MPEG-7 description tools onto an OWL ontology [15,17], but rather to use Semantic Web technologies to represent those constraints defined in natural language in the standard that cannot be expressed using XML Schema.

The paper is organized as follows. In the next section, we present a scenario motivating our work and justify the need for a partial formalization of MPEG-7. In section 3, we briefly introduce the notion of MPEG-7 profiles and we describe one of them as an example illustrating our work. In section 4, we detail how the MPEG-7 profiles can be formalized using Semantic Web technologies, building first an OWL ontology and rules capturing the semantic constraints, and developing then some tools converting from the XML-based MPEG-7 descriptions to RDF triples. In section 5, we illustrate the use of this work by describing an implemented Web-based application providing a semantic validation service for MPEG-7 multimedia descriptions. Finally, we give our conclusions and outline future work in section 6.

2   Motivating Scenario

The MPEG-7 description tools (Descriptors and Description Schemes) are represented using the *Description Definition Language* (DDL), the core part of the standard. Having collected high-level requirements [16] for the language, several DDL proposals have been proposed and evaluated [11]. A decision was taken to adopt the XML Schema recommendation [22] as the most appropriate schema language[1]. The MPEG-7 XML Schema defines numerous elements, types, and rules for their valid combinations. The standard, however, allows the specification of different descriptions with equivalent semantics. This raises interoperability problems when exchanging MPEG-7 descriptions, where different applications may use the standard differently.

---

[1] Several extensions (array and matrix datatypes) have been added in order to satisfy specific MPEG-7 requirements.

The following simple MPEG-7 description example illustrates the problem. The example is a format for reference data from the TREC Video Retrieval Evaluation, where the goal is to describe the shot structure of a video and the key frames representing each shot. The sample shown in Figure 1 is part of a description that validates against the MPEG-7 XML Schema. However, without additional knowledge about how MPEG-7 has been used, one cannot grasp the semantics of the elements in the description. For instance, the VideoSegment is used to represent at the same time the whole video content, the shots and the key frames.

```
<VideoSegment id="TRECVID2005_1">                <!-- whole video -->
  <MediaLocator>
    <MediaUri>20041116_110000_CCTV4_NEWS3_CHN.mpg</MediaUri>
  </MediaLocator>
  [...]
  <TemporalDecomposition gap="false" overlap="false">
    <VideoSegment id="shot1_1">                   <!--    shot     -->
      <MediaTime>
        <MediaTimePoint>T00:00:00:0F30000</MediaTimePoint>
        <MediaDuration>PT00H00M03S26116N30000F</MediaDuration>
      </MediaTime>
      <TemporalDecomposition>
        <VideoSegment id="shot1_1_RKF">      <!-- key frame -->
          <MediaTime>
            <MediaTimePoint>T00:00:01:27057F30000</MediaTimePoint>
          </MediaTime>
        </VideoSegment>
      </TemporalDecomposition>
    </VideoSegment>
    [...]
  </TemporalDecomposition>
</VideoSegment>
```

Fig. 1. A valid MPEG-7 shot structure description from the TRECVID evaluation

More generally, the problem comes from the modeling foundations of the languages used. XML Schema allows the derivation of new element types from existing ones, either by restriction or by extension. However, even if this derivation mechanism is reminiscent of the inheritance of object-oriented programming languages, it is rather a reuse of the content model defining the super type. The purpose of XML Schema is to define and to reuse the definitions of types and elements, that is, the set of syntactic constraints that define their content model, whereas well-defined semantics is the basis of the Semantic Web technologies. Coming back to our example, a proper OWL modeling of the VideoSegment should come with the definition of three different concepts rather than a single element type.

To overcome this interoperability issue, MPEG-7 has introduced the notion of profiles that constrain the use of subsets of the language, suitable for certain application domains (see section 3). For example, one can convert the previous MPEG-7 description into another one, conforming to the Detailed Audiovisual Profile (DAVP) [1], which specifies a number of semantic constraints in textual form. The conversion can be done by a straightforward XSLT transform that adjusts the structure of the description and adds the additional elements and attributes that are required by the DAVP schema. The result corresponding to Figure 1 is partly shown in Figure 2.

```
<VideoSegment id="TRECVID2005_1">
  <StructuralUnit
    href="urn:x-mpeg-7-davp:cs:StructuralUnitCS:2005:vis.programme"/>
  [...]
  <TemporalDecomposition gap="false" overlap="false"
    criteria="visual shots">
    <VideoSegment xsi:type="ShotType" id="shot1_1">
      <StructuralUnit
    href="urn:x-MPEG-7-davp:cs:StructuralUnitCS:2005:vis.shot"/>
      <MediaTime>
        <MediaTimePoint>T00:00:00:0F30000</MediaTimePoint>
        <MediaDuration>PT00H00M03S26116N30000F</MediaDuration>
      </MediaTime>
      <TemporalDecomposition criteria="key frames">
        <VideoSegment id="shot1_1_RKF">
          <StructuralUnit
    href="urn:x-MPEG-7-davp:cs:StructuralUnitCS:2005:vis.keyframe"/>
          <MediaTime>
            <MediaTimePoint>T00:00:01:27057F30000</MediaTimePoint>
          </MediaTime>
        </VideoSegment>
      </TemporalDecomposition>
    </VideoSegment>
    [...]
  </TemporalDecomposition>
</VideoSegment>
```

Fig. 2. A valid MPEG-7 shot structure description conforming to the DAVP profile

Additional elements include the StructuralUnit element on the segments that specify its semantic type (shot or key frame) and the criteria attribute of the decompositions. With these additional elements and attributes, it is then possible to distinguish between the different types of elements for which VideoSegment has been used. This enrichment of the XML schema allows more explicit expression of the structure of the description. The semantics of the DAVP constraints cannot, however, be formalized using XML Schema and thus cannot be checked for full semantic conformance. For example, the DAVP XML Schema requires the

`StructuralUnit` elements and criteria attributes to be present. But there is no way to check if their values are set correctly, for example, whether a key frame of a shot is a part of its temporal decomposition, or whether a key frame is not further decomposed into subparts. One can still put a `VideoSegment` element at any place allowed by the XML Schema and assign it the `StructuralUnit` value for key frame, or decompose a key frame further. The resulting description would still conform to the schema, although it would violate the non-formally represented semantic constraints of DAVP. Similarly, both the MPEG-7 and the DAVP XML schemas are able to specify whether a temporal decomposition of a video segment contains some *gap*, or whether the sub-segments *overlap*, by stating a true/false value for the corresponding XML attributes. It is not possible, however, to validate whether these values actually correspond to the time-coded temporal decomposition for a given description using only XML Schema conformance tools.

To summarize this discussion, we observe that the differences between the TRECVID MPEG-7 format and the DAVP MPEG-7 format prevent the exchange of the descriptions across applications, although the descriptions contain the same information. Worse, because of the lack of a formal definition of their semantics, it is not possible to automatically define mappings between the corresponding elements of the two descriptions. We thus propose a general method for formalizing the semantic constraints of any MPEG-7 profile.

3   MPEG-7 Profiling

Profiles have been proposed as a means of reducing the complexity of MPEG-7 descriptions. As specified in the standard, the definition of a profile consists of three parts, namely: *i) description tool selection*, i.e. the definition of the subset of description tools to be included in the profile, *ii) description tool constraints*, i.e. definition of constraints on the description tools such as restrictions on the cardinality of elements or on the use of attributes, and *iii) semantic constraints* that further describe the use of the description tools in the context of the profile. The first two parts are represented using the MPEG-7 DDL and result in a specific and more constrained XML Schema. The third part is currently expressed in natural language which prevents any automated process from efficiently tackling the complexity and interoperability problems associated with the use of MPEG-7 profiling for describing multimedia content.

3.1   Semantic Constraints

The semantic constraints in a profile can be compared to the prose descriptions specifying the semantics of any MPEG-7 descriptor or description scheme: both are written in natural language. Because of the generic nature of many description tools and the envisioned broad application area of MPEG-7, these descriptions do not contain many constraints with respect to their use. A profile is tailored to fit the needs of a much narrower application area, and thus the intended use of the description tools is better defined. The semantic constraints of the profile definition render the use of its description tools more precisely and thus avoid ambiguities in the descriptions.

The more generic and flexible a description tool is, the greater the need for semantic constraints. This is especially true for the structuring tools that allow the building of arbitrary hierarchical structures describing the media, spatial or temporal decompositions of the content into any kind of regions or segments. A profile can further specify the allowed structural decompositions for a given document type, genre or application. For example, a general VideoSegment descriptor describes some temporal segment of a video. In contrast, a VideoSegment used in the description structure of a profile might have a much more well-defined semantics. Hence, it can be refined in Shot, KeyFrame and Transition if the targeted application is to describe an automatic shot structure detection, or in Report, Interview and IndoorStudio in the context of the description of a weekly sports magazine broadcast on a TV channel [18].

3.2   Existing Profiles

Several profiles have been under consideration for standardization. Among them, the Simple Metadata Profile (SMP) allows the global definition of the general metadata of multimedia content of simple collections. The User Description Profile (UDP) consists of tools for describing user preferences and usage history in order to personalize multimedia content delivery. The Core Description Profile (CDP) contains the tools for the description of relationships between multimedia content, media information, creation information, usage information and semantic information. However, none of these adopted profiles include the visual and audio description tools that are useful, for example, for representing the results of feature extraction or automatic analysis of audiovisual content. Nor do they include the semantic constraints of the selected description tools that clarify how they should be used in the profile.

To overcome this flaw, a profile for the detailed description of single audiovisual content entities called Detailed Audiovisual Profile (DAVP)[2] has been proposed in [1]. This profile includes many of the MDS tools as well as those for audio and visual features description. Furthermore, it has been designed for improved support of interoperability between systems using MPEG-7, by avoiding possible syntax ambiguities and clarifying the use of the description tools in the profile. The semantic constraints thus play a crucial role in the DAVP definition since they define and constrain the use of the MPEG-7 description tools in the context of the profile. Because of the lack of formal semantics of XML Schema, these constraints are only described in textual form in the profile definition.

In summary, profiles are tailored towards specific application areas. They come with a schema that selects and further constrains a subset of description tools of the standard. Nevertheless, this does not solve the interoperability problem since the existing profiles are defined using the same informal notation as MPEG-7.

---

[2] http://mpeg-7.joanneum.at

4   Defining Semantics of MPEG-7 Descriptions

We detail in this section our approach to define the missing semantics layer on top of existing MPEG-7 profiles (section 4.1). We show that this task amounts to defining an ontology capturing the semantics of a given profile (section 4.2) as well as some additional rules to fully represent all its semantic constraints (section 4.3).

4.1   Overview of the Proposed Approach

MPEG-7 profiles are able to restrict the syntax variability of MPEG-7 descriptions (section 2). However, they cannot represent and thus check the correct use of the description tools according to their informal yet intended semantics. We therefore propose the following layered approach to validate *semantically* the conformance of MPEG-7 descriptions to a given profile:

Well-formedness. The MPEG-7 description is a well-formed XML document.
Validity. The description document validates against the full MPEG-7 XML Schema and the MPEG-7 profile it claims to conform to.
Semantics Consistency. The description is consistent with the ontology and rules describing the semantic constraints expressed informally in both the MPEG-7 standard and the profile definition.

We propose to use Semantic Web languages to express these constraints, and later inference tools to check the semantic consistency of the descriptions. The whole task can then be carried out with an appropriate combination of the following languages [10]:

- Using XML Schema to define largely structural constraints, i.e. what types are allowed and how they may be combined;
- Using OWL to formally capture the intended semantics of particular description tools;
- Using XSLT to convert from the original XML descriptions to RDF statements to assert the class-membership of particular description tools due to the presence of certain properties;
- Using additional rules to express relationships between structurally different but semantically equivalent constructs.

Therefore, achieving interoperability for MPEG-7 descriptions requires an ontology formally describing the profile (in terms of the concepts used) the MPEG-7 description purports to adhere to, and a way to convert automatically from the descriptions, the instances of the concepts modeled in this ontology. As an example, we show in section 4.2 how this method can be realized for the particular DAVP profile. The OWL expressivity being not enough for capturing all the semantics constraints, we give in section 4.3 some example rules to finalize the semantics formalization.

4.2 Building a Profile Ontology

Several attempts have been made to map the MPEG-7 description tools onto
an OWL ontology. An automatic mapping from XML Schema to OWL covering
the whole standard has been proposed [4,20]. However, without re-engineering
work, the resulting ontology is unable to capture the intended semantics, not
represented in the MPEG-7 schema, of the description tools. Other attempts
have manually modeled an MPEG-7 ontology. However, the result is either restricted to the upper level elements and types of MPEG-7 [9], or adapted to a
very specific use of the standard in a particular application [18]. Furthermore,
the additional semantic constraints specified in the profile definitions cannot be
taken into account.

Given this experience, we do not intend to completely map the MPEG-7 description tools onto an OWL ontology. We observe that there is also no need
to model the complete profile definitions in the ontology. We therefore only
capture the semantic constraints that cannot be handled by pure XML Schema
tools [2]. For example, for the DAVP profile, we model a DAVP ontology containing the concepts and properties necessary to represent some structural restrictions using a container-contained pattern (hasParent). In this ontology, both
GlobalTransition and Shot classes are defined as a subclass of a VideoSegment
with a particular value for the hasStructuralUnit property. The class KeyFrame
is also defined as a VideoSegment but one that cannot be further decomposed
temporally and that must be contained in a Shot. Figure 3 shows part of the resulting DAVP ontology for later describing a shot/keyframe structure compliant
with the DAVP profile.

One of the problems underlined in section 2 was that a single description tool
(e.g. VideoSegment) can be used in different contexts with a different semantics. It is now possible to model these different concepts in the ontology and to
automatically instantiate them from the XML descriptions using the additional
DAVP information. The functional role of a VideoSegment can be inferred by
an OWL reasoner given the value of its hasStructuralUnit property.

Finally, it has to be noted that some constraints implicitly defined in MPEG-7
are not formalized in the ontology. We detail in the next section how logical rules
can be used to capture these additional constraints.

4.3 Using Additional Rules for Expressing All Semantic Constraints

The limitations of XML Schema, especially with respect to MPEG-7, have been
discussed earlier in this paper. OWL itself has some serious limitations regarding the composition of properties [6,8]. The often cited example is the obvious
relationship between the composition of the isParentOf and isSiblingOf properties in conjunction with the class Male that yields the isUncleOf property:

$$\text{Male}(x) \sqcap \text{isSibling}(x,y) \sqcap \text{isParentOf}(y,z) \models \text{isUncleOf}(x,z)$$

Using rules in combination with Description Logics has been investigated for
quite a long time, thus a range of proposals exists [3,5,12]. In the Semantic

Web stack, it is expected that a rule language will complement the ontology layer. For example, the Semantic Web Rules Language (SWRL) [7] is a Horn clause rules extension to OWL for describing relationships between a composite property and other properties. Using SWRL would be a desirable approach, but was not feasible in practice since at the time of writing, no native SWRL-rules engine exists. The Jena Framework[3] is an example of an available integrated system combining an OWL reasoner with rules, and supporting both forward and backward reasoning. The rules are expressed in the N3 format, and the support of built-ins (validation, non-monocity) and RDF/OWL reasoning facilities are good arguments for its practical usage at the moment.

```
Class(a:Segment) Class(a:VideoSegment partial
  a:Segment
  restriction(a:hasStructuralUnit cardinality(1)))
Class(a:VideoProgrammeSegment complete
  a:VideoSegment
  restriction(a:hasStructuralUnit value (
    mpeg7:urn_x-MPEG-7-davp_cs_StructuralUnitCS_2005_vis.programme)))
Class(a:GlobalTransition complete
  a:VideoSegment
  restriction(a:hasParent allValuesFrom(a:VideoProgrammeSegment))
  restriction(a:hasStructuralUnit value (
    mpeg7:urn_x-MPEG-7-davp_cs_StructuralUnitCS_2005_vis.transition)))
Class(a:Shot complete
  a:VideoSegment
  restriction(a:hasKeyframe minCardinality(1))
  restriction(a:hasParent allValuesFrom(a:VideoProgrammeSegment))
  restriction(a:hasStructuralUnit value (
    mpeg7:urn_x-MPEG-7-davp_cs_StructuralUnitCS_2005_vis.shot))))
Class(a:Keyframe complete
  a:VideoSegment
  restriction(a:hasParent allValuesFrom(a:Shot))
  restriction(a:hasTemporalDecomposition maxCardinality(0))
  restriction(a:hasStructuralUnit value (
    mpeg7:urn_x-MPEG-7-davp_cs_StructuralUnitCS_2005_vis.keyframe))))
```

Fig. 3. Excerpt of the OWL ontology corresponding to the DAVP profile using an abstract syntax

Figure 4 shows part of the rule set used in our system to check for semantically invalid constructs in an MPEG-7 DAVP description. The first rule states that a KeyFrame without the required StructuralUnit descriptor will raise an error, while the second rule states that a KeyFrame must be contained in a Shot. When an invalid instance is found, it is annotated with a hasError property that can later be queried with SPARQL.

---

[3] http://jena.sourceforge.net/inference/ and
   http://jena.hpl.hp.com/juc2006/proceedings.html

```
# KeyFrame must have the appropriate structural unit
[keyframe_no_SU_rule:
        (?k a mpeg7:KeyFrame),
        noValue(?k mpeg7:hasStructuralUnit
         mpeg7:urn_x-mpeg-7-davp_cs_StructuralUnitCS_2005_vis.keyframe)
        ->
        (?k mpeg7:hasError mpeg7:DAVPViolationError)]

# KeyFrame must be contained in a Shot
[misplaced_keyframe_rule:
        (?k a mpeg7:KeyFrame),
        (?k mpeg7:hasParent ?s),
        noValue(?s a mpeg7:Shot)
        ->
        (?p mpeg7:hasError mpeg7:DAVPViolationError)]
```

Fig. 4. Example rules expressed in the Jena rule language

In the same way, the Jena rule language is used to represent the semantic constraints implicitly defined in MPEG-7 and that are not yet formalized in the ontology. For example, a rule can check whether the start/end time of the segments involved in a temporal decomposition are compatible with the start/end time range of their encompassing segment. It has to be noted that the use of rules comes potentially with serious scalability problems, since the languages mentioned above are known to be undecidable. However, during our experiments, we did not enter in the bad reasoning cases and the overall impact of these particular rules from the complexity point of view should be further explored in the future.

5 Implementation and Results

To illustrate the methodology detailed above, we have developed a Web-based application that provides a semantic validation service for any DAVP descriptions. The running application is available at
http://mpeg-7.joanneum.at/Validator/.
Figure 5 depicts the different components of the application.

- A first XSLT component takes an MPEG-7 document and pre-processes it to ensure that the input is compliant with respect to the DAVP schema (structural and XML-Schema-based syntactic validation);
- A second XSLT component takes the output of the first component and converts it into RDF-triples according to the DAVP OWL ontology;
- An inference service, such as the Jena Framework, takes as inputs the RDF-triples obtained previously and the additional logical rules modeling the semantic constraints of the DAVP profile in order to produce an inferred graph;

– Finally, a set of predefined SPARQL-queries can be executed on the resulting inferred graph in order to check the semantic validity of the MPEG-7 description.

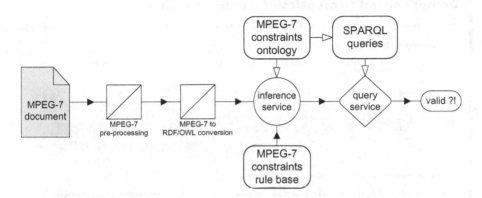

Fig. 5. Components and dataflow of the DAVP semantic validation application

Coming back to the scenario discussed in the section 2, we have pointed out that the problems caused by the unclear definitions of MPEG-7 description tools can be overcome by formalizing the semantic constraints defined in the profile. Hence, given an MPEG-7 description syntactically valid with respect to the DAVP schema, we want to validate semantically the description against the profile using the environment described above. We have shown in the previous sections how the semantic constraints related to the **Shot** and **KeyFrame** have been formalized. The top-level structure of the DAVP schema (the **Mpeg7** element and its children) has also been defined in the ontology[4].

According to the layered approach advocated in the section 4.1, the first step is to check the wellformedness and the syntactic validity of a description with respect to the DAVP schema. An XML processor such as Xerces is used in our application for this task. The XML description is then converted into an RDF graph that contains the instances of the classes defined in the DAVP ontology. An OWL reasoner such as RacerPro[5] or Pellet[6] is then used to check the consistency of the graph. Finally, the RDF representation of the description is processed by the Jena rule engine, and the resulting RDF graph is checked with a SPARQL query for any incorrect annotations produced by the inference engine. Figure 6 shows the user interface of our semantic validation web-based application.

Let's assume that an MPEG-7 document given as input to the semantic validation service has already been found to be valid against the DAVP XML Schema.

---

[4] The complete DAVP ontology as well as the rules modeled are available at
  http://mpeg-7.joanneum.at/Validator.

[5] http://www.racer-systems.com/

[6] http://www.mindswap.org/2003/pellet/

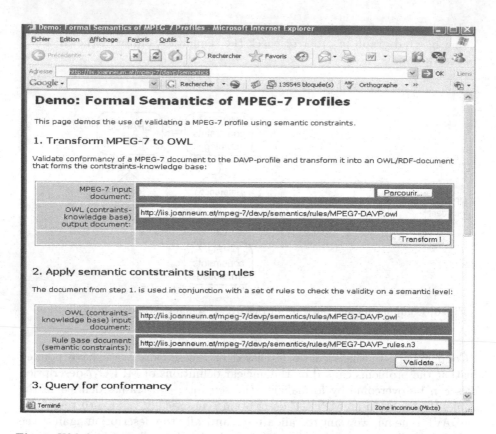

Fig. 6. Web-based application providing a semantic validation service for MPEG-7 descriptions

The following examples describe possible violations of the semantic constraints defined in the profile, while still being valid against the XML Schema:

– The OWL reasoner is able to detect "local" violations of the DAVP constraints such as a bad use of the `StructuralUnit` element on shots and key frames, or a key frame that is not contained in a shot description.
– An OWL reasoner cannot detect the missing `StructuralUnit` elements because of the open world assumption underlying to the OWL language. This error can be however detected by the Jena rule engine, thus motivating our choice to use additional rules for representing all semantic constraints of a profile.

Overall, this web-based application provides a semantic validation service for MPEG-7 descriptions. The times needed to perform the checks depends on the size and the complexity of the descriptions: it is quite fast (a few seconds) for some real MPEG-7 descriptions such as the ones produced by the TRECVid evaluation.

6   Conclusion and Future Work

In this paper, we have proposed an approach to overcome the interoperability problems that result from the lack of formal semantics of the MPEG-7 description tools by formalizing the semantic constraints of a profile. The approach is based on the definition of profiles, which are not just subsets of the MPEG-7 standard, but that also define a set of semantic constraints that specify the use of the description tools in a particular context. We have proposed to build an ontology that includes the concepts being described in the profile and to formalize the semantic constraints of the profile in terms of these concepts using description logic and rules. It was not our intention to completely represent the profile in the ontology, but only those constraints that cannot be validated using pure XML Schema tools. We have demonstrated the feasibility of this approach by modeling a subset of the DAVP profile, and we have set up a Web-based application that provides a semantic validation service for checking the conformance of any MPEG-7 descriptions against the semantic constraints of the profile.

As stated above, our approach requires the definition of semantic constraints on the description tools. One could argue that it is thus possible to apply this approach to the complete MPEG-7 standard, as there are many such constraints expressed in natural language in the standard. While this might work well for some of the description tools that have already a well-defined semantics, it is not the case for those generic Descriptors such as `VideoSegment`, which are the most problematic with respect to the interoperability problems discussed in Section 2. This means that a key prerequisite to make the proposed approach efficient is the definition of profiles that include sufficient semantic constraints to prevent ambiguities.

A next step in our work is to finalize the formalization of the semantic constraints for the whole DAVP profile. This task may require a stricter and more precise formulation of some of these constraints. There is of course always a trade-off between flexibility and strictness with respect to description tool semantics. If we require the semantic constraints to be very strict, this might prevent the use of any structures in the description not foreseen in the profile definition, even if they are used as extension and do not interfere with the structures defined in the profile. Thus it could be an option to introduce different levels of conformance to profile semantics, which one could name "semantic levels" of a profile.

Representing formally the semantic constraints of the MPEG-7 description tools is not only useful for semantically validating the descriptions as discussed in Section 5, but also for establishing mappings between profiles and heterogeneous MPEG-7 descriptions. Actually, the greatest potential with semantic definitions of MPEG-7 profiles is in the ability to use these descriptions to relate the content to other audiovisual streams/segments described using alternative MPEG-7 profiles or other domain ontologies. Current multimedia applications on the Web, such as multimedia search engines, need to index multimedia metadata from heterogeneous sources. Formalizing the semantics of the profiles used for representing this metadata allows the auto mapping of the descriptions, guess the semantics of each audiovisual segments, etc. In the future, we plan to investigate further how the approach presented in this paper can be used in this particular use case.

Acknowledgments

The authors would like to thank the following colleagues at JOANNEUM RE-SEARCH (Georg Thallinger, Herwig Zeiner and Peter Schallauer) and at CWI (Lynda Hardman, Jacco van Ossenbruggen) for their feedback on the prototype and earlier versions of this paper. The research leading to this paper was partially supported by the European Commission under contract FP6-027026, "Knowledge Space of semantic inference for automatic annotation and retrieval of multimedia content - K-Space", and under contract IST-2-511316, "IP-RACINE: Integrated Project - Research Area CINE".

References

1. Werner Bailer and Peter Schallauer. The Detailed Audiovisual Profile: Enabling Interoperability between MPEG-7 based Systems. In 12th International MultiMedia Modelling Conference (MMM'06), pages 217–224, Beijing, China, 2006.
2. Werner Bailer, Peter Schallauer, Michael Hausenblas, and Georg Thallinger. MPEG-7 Based Description Infrastructure for an Audiovisual Content Analysis and Retrieval System. In Proceedings of SPIE - Storage and Retrieval Methods and Applications for Multimedia, pages 284–295, San Jose, California, USA, 2005.
3. Francesco M. Donini, Maurizio Lenzerini, Daniele Nardi, and Andrea Schaerf. AL-log: Integrating Datalog and Description Logics. Journal of Intelligent Information Systems, 10(3):227–252, 1998.
4. Roberto Garcia and Oscar Celma. Semantic Integration and Retrieval of Multimedia Metadata. In 5th International Workshop on Knowledge Markup and Semantic Annotation (SemAnnot'05), Galway, Ireland, 2005.
5. Benjamin Grosof, Ian Horrocks, Raphael Volz, and Stefan Decker. Description Logic Programs: Combining Logic Programs with Description Logics. In 12th International World Wide Web Conference (WWW'03), Budapest, Hungary, 2003.
6. Ian Horrocks, Peter F. Patel-Schneider, Sean Bechhofer, and Dmitry Tsarkov. OWL rules: A proposal and prototype implementation. Journal of Web Semantics, 3(1):23–40, 2005.
7. Ian Horrocks, Peter F. Patel-Schneider, Harold Boley, Said Tabet, Benjamin Grosof, and Mike Dean. SWRL: A Semantic Web Rule Language Combining OWL and RuleML. Technical report, World Wide Web Consortium, May 2004. http://www.w3.org/Submission/SWRL/.
8. Ian Horrocks, Peter F. Patel-Schneider, and Frank van Harmelen. From $\mathcal{SHIQ}$ and RDF to OWL: The making of a web ontology language. Journal of Web Semantics, 1(1):7–26, 2003.
9. Jane Hunter. Adding Multimedia to the Semantic Web - Building an MPEG-7 Ontology. In First International Semantic Web Working Symposium (SWWS'01), Stanford, California, USA, 2001.

10. Jane Hunter and Carl Lagoze. Combining RDF and XML Schemas to Enhance Interoperability Between Metadata Application Profiles. In 10th *International World Wide Web Conference (WWW'01)*, pages 457–466, Hong Kong, 2001.
11. Jane Hunter and Frank Nack. An overview of the MPEG-7 Description Definition Language (DDL) proposals. *Signal Processing: Image Communication*, 16(1-2):271–293, 2000.
12. Alon Y. Levy and Marie-Christine Rousset. Combining horn rules and description logics in carin. *Artificial Intelligence*, 104(1-2):165–209, 1998.
13. MPEG-7. Multimedia Content Description Interface. Standard No. ISO/IEC n°15938, 2001.
14. Frank Nack, Jacco van Ossenbruggen, and Lynda Hardman. That Obscure Object of Desire: Multimedia Metadata on the Web (Part II). *IEEE Multimedia*, 12(1), 2005.
15. OWL. Web Ontology Language Reference. W3C Recommendation, 10 February 2004. http://www.w3.org/TR/owl-ref/.
16. Fernando Pereira. MPEG-7 Requirements Document V.16. ISO/IEC JTC1/SC29/WG11/N4510. Pattaya, Thailand, December 2001.
17. RDF. Ressource Description Framework Primer. W3C Recommendation, 10 February 2004. http://www.w3.org/TR/rdf-primer/.
18. Raphaël Troncy. Integrating Structure and Semantics into Audio-visual Documents. In 2nd *International Semantic Web Conference (ISWC'03)*, pages 566–581, Sanibel Island, Florida, USA, 2003.
19. Raphaël Troncy and Jean Carrive. A Reduced Yet Extensible Audio-Visual Description Language: How to Escape From the MPEG-7 Bottleneck. In 4th *ACM Symposium on Document Engineering (DocEng'04)*, Milwaukee, Wisconsin, USA, 2004.
20. Chrisa Tsinaraki, Panagiotis Polydoros, and Stavros Christodoulakis. Interoperability support for Ontology-based Video Retrieval Applications. In 3rd *International Conference on Image and Video Retrieval (CIVR'04)*, Dublin, Ireland, 2004.
21. Jacco van Ossenbruggen, Frank Nack, and Lynda Hardman. That Obscure Object of Desire: Multimedia Metadata on the Web (Part I). *IEEE Multimedia*, 11(4), 2004.
22. XML Schema. W3C Recommendation, 2 May 2001. http://www.w3.org/XML/Schema.

Trend Detection in Folksonomies

Andreas Hotho[1], Robert Jäschke[1,2], Christoph Schmitz[1], and Gerd Stumme[1,2]

[1] Knowledge & Data Engineering Group, Department of Mathematics and Computer Science,
University of Kassel, Wilhelmshöher Allee 73, D–34121 Kassel, Germany
http://www.kde.cs.uni-kassel.de
[2] Research Center L3S, Expo Plaza 1, D–30539 Hannover, Germany
http://www.l3s.de

Abstract. As the number of resources on the web exceeds by far the number of documents one can track, it becomes increasingly difficult to remain up to date on ones own areas of interest. The problem becomes more severe with the increasing fraction of multimedia data, from which it is difficult to extract some conceptual description of their contents.

One way to overcome this problem are social bookmark tools, which are rapidly emerging on the web. In such systems, users are setting up lightweight conceptual structures called folksonomies, and overcome thus the knowledge acquisition bottleneck. As more and more people participate in the effort, the use of a common vocabulary becomes more and more stable. We present an approach for discovering topic-specific trends within folksonomies. It is based on a differential adaptation of the PageRank algorithm to the triadic hypergraph structure of a folksonomy. The approach allows for any kind of data, as it does not rely on the internal structure of the documents. In particular, this allows to consider different data types in the same analysis step. We run experiments on a large-scale real-world snapshot of a social bookmarking system.

1 Social Resource Sharing and Folksonomies

With the growth of the web, both the number and the heterogeneity of types of available resources have increased dramatically. The management of such a collection of resources includes many subtasks like search, retrieval, clustering, reasoning, and knowledge discovery. For all these tasks, some sort of conceptual description of the documents is essential. While there are many approaches that have been applied successfully for years for extracting such descriptions from text documents — ranging from the bag-of-words model for information retrieval to ontology learning — there are fewer solutions for images, videos, audio tracks and music data up to now. The way from the features of the different resources to a conceptual description is generally far more difficult for multimedia data. Furthermore, these techniques have to be developed separately for each kind of data. For applications like the detection of trends from a collection of resources consisting of several types of (multimedia) data — which is the topic of this paper — first a common format for the representation of the conceptual model plus extraction techniques for each of the data types would have to be defined.

Complementing the extraction of conceptual descriptions from the documents themselves, social resource sharing tools are currently emerging on the web, as a part of what is called "social software" or "Web 2.0". In these user-centric publishing and

Y. Avrithis et al. (Eds.): SAMT 2006, LNCS 4306, pp. 56–70, 2006.
© Springer-Verlag Berlin Heidelberg 2006

knowledge management platforms, a conceptual description is provided to each document by the user in the form of a collection of 'tags', i.e., of arbitrary, user-defined catchwords. As this description is independent of the format of the resource, the social tagging approach provides a unified model for all kinds of resources, including in particular multimedia formats.

Social resource sharing tools, such as Flickr[1] or del.icio.us[2] (see Fig. 1), have acquired large numbers of users within less than two years. The social photo gallery Flickr, for instance, is estimated to have over a million users. The reason for the immediate success of these systems is the fact that no specific skills are needed for participating, and that these tools yield immediate benefit for each individual user (e.g. organizing ones bookmarks in a browser-independent, persistent fashion) without too much overhead. Large numbers of users have created huge amounts of information within a very short period of time. The frequent use of these systems shows clearly that web- and folksonomy-based approaches are able to overcome the knowledge acquisition bottleneck, which was a serious handicap for many knowledge-based systems in the past.

Social resource sharing systems are web-based systems that allow users to upload their resources, and to label them. All these systems share the same core functionality. Once a user is logged in, he can add a resource to the system, and assign arbitrary labels, so-called *tags*, to it. Resources can be almost anything. In systems such as our *BibSonomy*,[3] for instance, resources are bookmarks and bibliographic references, in *Flickr* they are photos, in *last.fm*[4] music files, in *YouTube*[5] videos, and in *43Things*[6] even goals in private life.

The collection of all assignments of a user is called his *personomy*, the collection of all personomies is called *folksonomy*. The user can also explore the personomies of the other users in all dimensions: for a given user he can see the resources that user had uploaded, together with the tags he had assigned to them; when clicking on a resource he sees which other users have uploaded this resource and how they tagged it; and when clicking on a tag he sees who assigned it to which resources (see Fig. 1).

The word 'folksonomy' is a blend of the words 'taxonomy' and 'folk', and stands for conceptual structures created by the people. Folksonomies are thus a bottom-up complement to more formalized Semantic Web technologies, as they rely on *emergent semantics* [17,18] which result from the converging use of the same vocabulary.

In this paper, we will analyze this emergence of common semantics by exploring trends in the folksonomy. Since the structure of a folksonomy is symmetric with respect to the dimensions 'user', 'tag', and 'resource', we can apply the same approach to study upcoming users, upcoming tags, and upcoming resources. We present a technique for analyzing the evolution of topic-specific trends. Our approach is based on our *FolkRank* algorithm [10], a differential adaptation of the PageRank algorithm [3] to the tri-partite hypergraph structure of a folksonomy. Compared to pure co-occurrence

---

[1] http://www.flickr.com/
[2] http://del.icio.us
[3] http://www.bibsonomy.org
[4] http://www.last.fm
[5] http://www.youtube.com/
[6] http://www.43things.com/

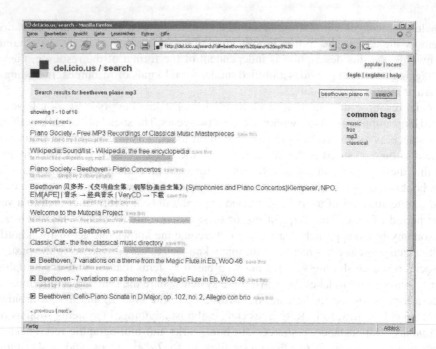

Fig. 1. Del.icio.us, a popular social bookmarking system

counting, FolkRank takes also into account elements that are related to the focus of interest with respect to the underlying graph/folksonomy. In particular, FolkRank ranks synonyms higher, which usually do not occur in the same bookmark posting together.

With FolkRank, we compute topic-specific rankings on users, tags, and resources. In a second step, we can then compare these rankings for snapshots of the system at different points in time. We can discover both the absolute rankings (who is in the Top Ten?) and winners and losers (who rose/fell most?).

The contributions of this work are:

Ranking in folksonomies. We describe a general ranking scheme for folksonomy data. The scheme allows in particular for topic-specific ranking.

Trend detection. We introduce a trend detection measure which allows to determine which tags, users, or resources have been gaining or losing in popularity in a given time interval. Again, this measure allows to focus on specific topics.

Application to arbitrary folksonomy data. As the ranking is solely based on the graph structure of the folksonomy – which is resource-independent – we can also apply it to any kind of resources, including in particular multimedia objects, but also office documents which typically do not have a hyperlink structure per se. It can even be applied to an arbitrary mixture of these content types. Actually, the content of the tagged resources will not have to be accessible in order to manage them in a folksonomy system.

Evaluation. We have applied our method to a large-scale dataset from an actual folksonomy system.

The paper is organized as follows. In the next section, we describe our ranking and trend detection approach. In Section 3, we apply the approach to a large-scale dataset, a one-year snapshot of the del.icio.us system. Section 4 discusses related work, and Section 5 concludes with an outlook on future topics in this field.

2 Trend Detection in Folksonomies

For discovering trends in a social resource sharing system, we will need snapshots of its folksonomy at different points of time. For each snapshot, we will need a ranking, such that we can compare the rankings of consecutive snapshots. As we also want to discover topic-specific trends, we will additionally need a ranking method that allows to focus on the specific topic. We will make use of our search and ranking algorithm *FolkRank* [10] which we summarize below.

2.1 Basic Notions

A folksonomy basically describes the users, resources, tags, and allows users to assign (arbitrary) tags to resources. We will make use of the following notions. A *folksonomy* is a tuple $\mathbb{F} := (U, T, R, Y, \prec)$ where

- U, T, and R are finite sets, whose elements are called *users*, *tags* and *resources*, resp.,
- Y is a ternary relation between them, i. e., $Y \subseteq U \times T \times R$, whose elements are called tag assignments (TAS for short), and
- $\prec$ is a user-specific subtag/supertag-relation, i. e., $\prec \subseteq U \times T \times T$, called *subtag/supertag relation*.

The *personomy* $\mathbb{P}_u$ of a given user $u \in U$ is the restriction of $\mathbb{F}$ to u, i. e., $\mathbb{P}_u := (T_u, R_u, I_u, \prec_u)$ with $I_u := \{(t,r) \in T \times R \mid (u,t,r) \in Y\}$, $T_u := \pi_1(I_u)$, $R_u := \pi_2(I_u)$, and $\prec_u := \{(t_1, t_2) \in T \times T \mid (u, t_1, t_2) \in \prec\}$ (whereas π stands for the projection).

Users are typically described by their user ID, and tags may be arbitrary strings. What is considered as a resource depends on the type of system. For instance, in del.icio.us, the resources are URLs, and in Flickr, the resources are pictures. From an implementation point of view, resources are internally represented by some ID.

In this paper, we do not make use of the subtag/supertag relation for sake of simplicity. I. e., $\prec = \emptyset$, and we will simply note a folksonomy as a quadruple $\mathbb{F} := (U, T, R, Y)$. This structure is known in Formal Concept Analysis [20,7] as a *triadic context* [13,19]. An equivalent view on folksonomy data is that of a tripartite (undirected) hypergraph $G = (V, E)$, where $V = U \dot\cup T \dot\cup R$ is the set of nodes, and $E = \{\{u, t, r\} \mid (u, t, r) \in Y\}$ is the set of hyperedges ($\dot\cup$ is the disjunctive union).

2.2 Ranking

In this section we recall the principles of the *FolkRank* algorithm that we developed for supporting Google-like search in folksonomy-based systems. It is inspired by the seminal PageRank algorithm [3].

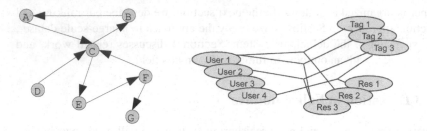

Fig. 2. Webgraph vs. Folksonomy Hypergraph

Because of the different nature of folksonomies compared to the web graph (undirected triadic hyperedges instead of directed binary edges, see Figure 2), PageRank cannot be applied directly on folksonomies. In order to employ a weight-spreading ranking scheme on folksonomies, we will overcome this problem in two steps. First, we transform the hypergraph into an undirected graph. Then we apply a differential ranking approach that deals with the skewed structure of the network and the undirectedness of folksonomies.

Folksonomy-Adapted Pagerank. First we convert the folksonomy $\mathbb{F} = (U, T, R, Y)$ into an *un*directed tri-partite graph $G_{\mathbb{F}} = (V, E)$ as follows.

1. The set V of nodes of the graph consists of the disjoint union of the sets of tags, users and resources: $V := U \dot\cup T \dot\cup R$. (The tripartite structure of the graph can be exploited later for an efficient storage of the adjacency matrix and the implementation of the weight-spreading iteration in the FolkRank algorithm.)
2. All co-occurrences of tags and users, users and resources, tags and resources become edges between the respective nodes: $E := \{\{u, t\} \mid \exists r \in R : (u, t, r) \in Y\} \cup \{\{t, r\} \mid \exists u \in U : (u, t, r) \in Y\} \cup \{\{u, r\} \mid \exists t \in T : (u, t, r) \in Y\}$.

The original formulation of PageRank [3] reflects the idea that a page is important if there many pages linking to it, and if those pages are important themselves. The distribution of weights can thus be described as the fixed point of a weight passing scheme on the web graph. This idea was extended in a similar fashion to bipartite subgraphs of the web in HITS [12] and to n-ary directed graphs in [21]. We employ the same underlying principle for our ranking scheme in folksonomies. The basic notion is that a resource which is tagged with important tags by important users becomes important itself. The same holds, symmetrically, for tags and users, thus we have a graph of vertices which are mutually reinforcing each other by spreading their weights.

Like PageRank, we employ the random surfer model, a notion of importance for web pages that is based on the idea that an idealized random web surfer normally follows hyperlinks, but from time to time jumps to a new webpage without following a link. This results in the following definition. The rank of the vertices of the graph are the entries in the fixed point $\vec{w}$ of the weight spreading computation

$$\vec{w} \leftarrow dA\vec{w} + (1 - d)\vec{p} \ , \tag{1}$$

where $\vec{w}$ is a weight vector with one entry for each web page, A is a row-stochastic version of the adjacency matrix of the graph $G_{\mathbb{F}}$ defined above, $\vec{p}$ is the random surfer component that outweighs the loss of weight in dangling links, and $d \in [0, 1]$ is determining the influence of $\vec{p}$. Usually, one will choose $\vec{p} = \mathbf{1}$, i.e., the vector composed by 1's, to achieve uniform damping. In order to compute personalized PageRanks, however, $\vec{p}$ can be used to express user preferences by giving a higher weight to the components which represent the user's preferred web pages. If $||\vec{w}||_1 = ||\vec{p}||_1$,[7] the weight in the system will remain constant.

As the graph $G_{\mathbb{F}}$ is undirected, most of the weight that went through an edge at moment t will flow back at $t + 1$. The results are thus rather similar (but not identical, due to the damping) to a ranking that is simply based on edge degrees. The reason for applying the more expensive PageRank approach nonetheless is that its random surfer vector allows for topic-specific ranking.

FolkRank — Topic-Specific Ranking. As the graph $G_{\mathbb{F}}$ that we created in the previous step is undirected, we face the problem that an application of the original PageRank would result in weights that flow in one direction of an edge and then 'swash back' along the same edge in the next iteration, so that one would basically rank the nodes in the folksonomy by their degree distribution. This makes it very difficult for other nodes than those with high edge degree to become highly ranked, no matter what the preference vector is.

This problem is solved by the *differential* approach in FolkRank, which computes a personalized ranking of the elements in a folksonomy as follows:

1. The preference vector $\vec{p}$ is used to determine the topic. It may have any distribution of weights, as long as $||\vec{w}||_1 = ||\vec{p}||_1$ holds. Typically a single entry or a small set of entries is set to a higher value, and the remaining weight is equally distributed over the other entries. Since the structure of folksonomies is symmetric, we can define a topic by giving a higher value to either one or more tags and/or one or more users and/or one or more resources.
2. Let $\vec{w_0}$ be the fixed point from Equation (1) with $d = 1$.
3. Let $\vec{w_1}$ be the fixed point from Equation (1) with $d < 1$. In our experiments, we set $d = 0.85$.
4. $\vec{w} := \vec{w_1} - \vec{w_0}$ is the final weight vector.

Thus, we compute the winners and losers of the mutual reinforcement of nodes when a user preference is given, compared to the baseline without a preference vector. We call the resulting weight $\vec{w}[x]$ of an element x of the folksonomy the *FolkRank* of x. In [10] we showed that $\vec{w}$ provides indeed valuable results on a large-scale real-world dataset while $\vec{w_1}$ provides an unstructured mix of topic-relevant elements with elements having high edge degree.

2.3 Trend Detection

In order to analyze the trends around a specific topic, we first have to describe the topic by defining the preference vector $\vec{p}$. Then we compute, for each point in time

---

[7] . . . and if there are no rank sinks – but this holds trivially in our graph $G_{\mathbb{F}}$.

$t \in \{0, \dots, n\}$, the rank vector $\vec{w}_t$ within the folksonomy $\mathbb{F}_t$ which consists of all tag assignments performed before t.[8]

We select then from the resulting rank vectors those entries which are assigned to one of the three dimensions 'tags', 'users', and 'resources' — depending on where we want to see rising and falling elements. Else an analysis would be difficult, since users have higher weights than tags, which in their turn have higher weights than resources, due to the different sizes of the sets U, T, and R.

As the total weight in the system will differ at different points of time because of new tags, users, and resources, we normalize at last each rank vector such that its largest value equals 1. This allows to compare rankings from different points in time. If the preference vector has only one distinguished element, then this element is the one with the highest value in the resulting weight vector. The closer another entry is to this value, the more important is its associated element to the topic. By plotting the values of the Top 10 or Top 20 over time, one can thus discover the rise and fall of the most popular elements. Figure 3 shows such a plot for the del.icio.us users which are most important for the topic 'music', while Figure 4 shows the tags which are most important for the topic 'politics'. How these diagrams are to be read, and what the most important findings are, will be described in detail in the next section.

Going a step further, we may not only be interested in the most important elements, but also in those where the increase or decrease of rank is the steepest. To this end, we have developed the following *popularity change* measure, which allows for detecting topic-specific trends.

Assume x is a tag, user or resource of the folksonomy $\mathbb{F}$, i. e. $x \in U \cup T \cup R$. (In the following, we assume it is a resource; the same methods apply symmetrically for tags and users.) Similar to the relative change used for word occurrences in [11], we define the *popularity change* $pc_{t_0 \rightarrow t_1}(x)$ of x from t_0 to t_1 as follows.

At times $t_0 < t_1$, let the resource x be ranked at position r_0 and r_1, respectively, in the descending weight order of the FolkRank computation. Let n_0 and n_1 be the sizes $|R|$ of the resource dimension at times t_0, t_1. The popularity change is defined as

$$pc_{t_0 \rightarrow t_1}(x) := \left(\frac{r_0}{n_0} - \frac{r_1}{n_1} \right) \log_{10} \left(\frac{n_1}{r_1} \right) \qquad (2)$$

(where elements not present at time t_i are treated as being positioned at $r_i = n_i + 1$). Here, the fractions in the first term indicate the relative positions of x at the given times, $1/n_i$ being the best (i. e. having maximum FolkRank) and 1 being the worst. The second term discounts the change with respect to the relative position where the change took place: to get from a top 90 % position to a top 80 % one would be considered three times easier than to get from the top 0.09 % to the top 0.08 %.

Combined with a topic-directed FolkRank computation, we use this measure of a change in popularity to get an insight into what are the trends in a certain community in the folksonomy. We point out the winning and losing elements of the folksonomy in a given time interval.

---

[8] If no entries were deleted, $\mathbb{F}_{t+1}$ contains thus $\mathbb{F}_t$, for all t.

3 Experiments

3.1 Evaluation of Popularity Change in del.icio.us

In order to evaluate our approach, we have analyzed the popular social bookmarking sytem del.icio.us.[9] Del.icio.us is a server-based system with a simple-to-use interface that allows users to organize and share bookmarks on the internet. The resources del.icio.us is pointing to cover various formats (text, audio, video, etc.). In particular, the system is not restricted to a single type (like photos in Flickr). As discussed above, our approach is specially suited for this situation. In addition to the URL, del.icio.us allows to store a description, an extended description, and tags (i. e., arbitrary labels). Del.icio.us is online for a sufficiently long time (since May 2002) to allow for extracting significant time series.

For our experiments, we collected data from the del.icio.us system between July 27 and July 30, 2005 in the following way. Initially we used wget starting from the start page of del.icio.us to obtain nearly 6900 users and 700 tags as a starting set. Out of this dataset we extracted all users and resources (i. e., del.icio.us' MD5-hashed urls). We downloaded in a recursive manner user pages to get new resources and resource pages to get new users. Furthermore we monitored the del.icio.us start page to gather additional users and resources. This way we collected a list of several thousand usernames which we used for accessing the first 10000 resources each user had tagged. From the collected data we finally took the user files to extract resources, tags, dates, descriptions, extended descriptions, and the corresponding username.

We obtained a folksonomy with $|U| = 75,242$ users, $|T| = 533,191$ tags and $|R| = 3,158,297$ resources, related by in total $|Y| = 17,362,212$ tag assignments. We created monthly snapshots as follows. $\mathbb{F}_0$ contains all tag assignments performed on or before June 15, 2004, together with the involved users, tags, and resources; $\mathbb{F}_1$ all tag assignments performed on or before July 15, 2004, together with the involved users, tags, and resources; and so on until $\mathbb{F}_{13}$ which contains all tag assignments performed on or before July 15, 2005, together with the involved tags, users, and resources.

Figure 3 shows the evolution of the ranking of all users tags that were among the Top 10 in at least one month for the topic 'music'. The diagram was obtained with $d = 0.85$, and the preference vector $\vec{p}$ set such that the tag 'music' gets 50% of the overall preference, the rest is spread uniformly as described above. The user names have been omitted for privacy reasons. The diagram shows three outstanding users. The first one could keep the top position for the first four months, followed by a steep fall. Another user could approach him steadily during the first four months, followed by almost the same fall. The fall of both was caused by the steep rise of a new user, which also shadowed the rankings of all other users related to 'music'. A detailed analysis of this user's data in the system revealed us that he posted more than 5500 bookmarks, 85 % of which tagged with 'music'. In total he used only about 100 tags. The 5500 bookmarks account for about 2% of *all* occurrences of 'music' in the system (with more than 70.000 users in the system at that time), and are about 3.5 times as many as those of the second user for that tag.

---

[9] http://del.icio.us

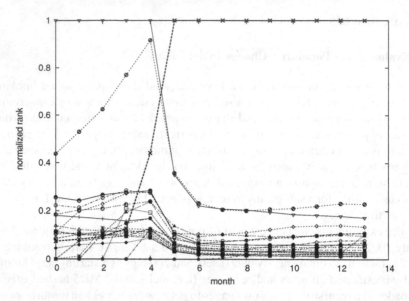

Fig. 3. Evolution of the ranking of users related to 'music'. User names are omitted for privacy reasons.

Figure 4 shows the evolution of all tags that were among the Top Ten in at least one month for the topic 'politics'. The line for the topic 'politics' itself can't be seen, as it has a rank of 1. The diagram was obtained with $d = 0.85$, and the preference vector $\vec{p}$ set such that the tag 'politics' gets 50% of the overall preference, while the rest is spread uniformly over the other tags, users and resources. The diagram shows that the early users of del.icio.us were more critical/idealistic, as they used tags like 'activism', 'humor', 'war', and 'bushco'[10]. With increasing time, the popularity of these tags faded, and the tags turned to a more uniform distribution, as the closing lines at the right of the figure indicate. In particular one can discover the rise of the tags 'bush' and 'election', both having a peak around the election day, November 2nd, 2004, and remaining on a high level afterwards. Within the analysis of the topic 'technology' (not displayed due to space reasons), we have discovered a similar trend: The early adoptors of del.icio.us used the tag 'technology' together with tags like 'culture', 'society' or 'apple', while later tags like 'gadgets', 'news' or 'future' rise, converging towards more mainstream topics.

Both Figures 3 and 4 show that there is a change of structure in autumn 2004 (month 4 in the diagrams). This is supported by Figure 5 showing the development of the top resources. Analysing possible reasons for this change in behavior, one indicator is that the number of elements passed in month 4 the threshold of 10.000 users, 70.000 tags, and 500.000 resources. Apparently, with this number of users, one reaches a critical mass which modifies the inherent behavior of such a system. Figure 5 shows the rank of those resources which where among the top 5 at the beginning or the end. Our hypothesis that the del.icio.us community changes significantly at month 4 is supported

---

[10] In del.icio.us, 'bushco' was used for tagging webpages about the interference of politics and economics in the U. S. administration.

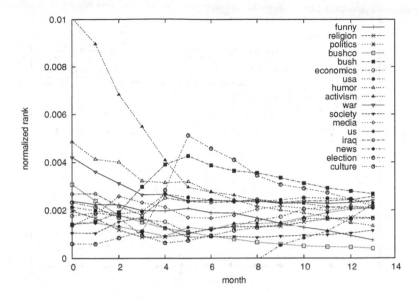

Fig. 4. Evolution of the ranking of tags related to 'politics' over time. 'Politics' has value 1.0 due to normalization and is left out for clarity of the presentation of the other values.

by two observations: specific topics of the beginning, such as web design, see a decline, while on the other hand, mainstream pages gain rapidly, such as Slashdot, as well as pages concerned with folksonomies per se.

Finally, we have analyzed those resources that were the strongest winners and losers within specific topics, according to the popularity change measure defined in Section 2.3, to automatically identify trends within certain topics in del.icio.us. Our aim was to discover trends in the Semantic Web community in the month around the European Semantic Web Conference (ESWC) 2005.

For the computation, we took those resources that ended up in the Top 100 in the June 2005 ranking for the preference vector highlighting the tags 'semantic_web', 'semantic', 'web', and 'semanticweb', since the top results e. g. in searches are typically the ones attracting the most users. For these 100 URLs, we computed the popularity change coefficient from May 15 to June 15. Table 1 shows the 20 ULRs with the highest popularity change.

The top winner (#1) is a site about shallow semantic markup in XHTML, which was obviously first discovered by the community during the period under consideration and made it to the 39th position out of 2.2M resources; the corresponding line in Table 1 shows that the FolkRank value and the position in May is undefined for this resource, while the rank in June is 0.13065 and the position in this ranking is 39. Among the followers are articles about the Semantic Web and folksonomies (e. g. #2, #3, #5, #8, #9, #16), pages about new Semantic Web projects (#4, #15, #17, #19), or events such as the Scripting workshop (#7) that took place together with the ESWC conference during the period under consideration, introducing new Semantic Web projects. Note that while the #1 page leaped from nowhere to the 39th position out of 2.2 million

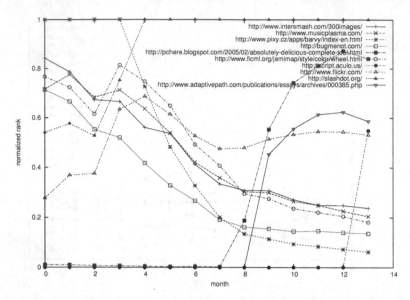

Fig. 5. Evolution of the global ranking of resources, without specific preference vector

entries, the popularity change measure still honors movements at the top of the ranking: the Piggy Bank site (which is an important semantic web project that has been promoted at the ESWC conference), improving from 21 to 1 in the period, still gets into the top 15 winners.

Together, the results of the FolkRank computation and the popularity change measure presented in this section can thus be used to get an insight into the structure and development of communities in folksonomy systems, independent of and across different media types.

3.2   Comparison with the Interestingness of Dubinko et. al.

Closest to the approach of this paper is the visualization of Dubinko et. al. [6]. We tried to get an insight into how our FolkRank compares to the interestingness of [6]. In that paper, the authors introduce an efficient way of mining large-scale folksonomy data sets for frequent tags in given time intervals. A measure of *interestingness*, is introduced and computed for a sliding one-day window over a Flickr dataset. Similar to the TF/IDF measure from Information Retrieval, the interestingness is defined as $Int(o, I) = \sum_{i \in I} \gamma(o, i)/(C + \gamma(o))$, where $\gamma(o, i)$ is the number of occurrences of object o in time interval i out of a larger interval I, and $\gamma(o)$ is the total number of occurrences of o. As the interestingness is based on a count of occurrences of items[11] in a given interval, it does not allow for an easy integration of topic-specific rankings. Thus, one obtains a ranking of one particular tag (user, resource), which does not generalize to related elements of the folksonomy.

---

[11] In [6], only tags are evaluated. Still, the method can be applied symmetrically to users and resources.

Table 1. Popularity Change from May 15 to June 15, 2005

#	URL	Pop.Chg.	May Rank	May Pos	June Rank	June Pos
1	http://mezzoblue.com/downloads/markupguide/	4.822604	undef	undef	0.13065	39
2	http://www.betaversion.org/~stefano/linotype/news/89/	4.515983	undef	undef	0.08296	79
3	http://shirky.com/writings/ontology_overrated.html	0.073704	0.00866	28598	0.39329	4
4	http://simile.mit.edu/piggy-bank/index.html	0.000805	0.05160	377	0.18740	24
5	http://www.dlib.org/dlib/april05/hammond/04hammond.html	0.000183	0.08831	142	0.09532	61
6	http://www.w3.org/2004/02/skos/	0.000175	0.08282	155	0.08369	78
7	http://www.semanticscripting.org/SFSW2005/	0.000134	0.09427	124	0.09055	67
8	http://www.scientificamerican.com/article.cfm?...	0.000133	0.08396	152	0.07208	97
9	http://jena.hpl.hp.com/~stecay/papers/xmleur...	0.000129	0.09979	111	0.09990	56
10	http://www.tantek.com/presentations/2004ete...	0.000112	0.09047	137	0.07407	92
11	http://users.bestweb.net/~sowa/peirce/ontometa.htm	0.000111	0.10273	106	0.09550	60
12	http://www.sciam.com/print_version.cfm?articleID=...	0.000101	0.09608	121	0.08178	81
13	http://www.xml.com/pub/a/2001/01/24/rdf.html	0.000089	0.09391	127	0.07314	94
14	http://developers.technorati.com/wiki/hCalendar	0.000071	0.09748	117	0.07389	93
15	http://simile.mit.edu/piggy-bank/	0.000057	0.29151	21	1.00000	1
16	http://en.wikipedia.org/wiki/Semantic_web	0.000033	0.10472	102	0.07186	98
17	http://www.semanticplanet.com/	0.000025	0.10893	91	0.07510	90
18	http://pchere.blogspot.com/2005/02/absolute...	0.000023	0.18154	41	0.13729	35
19	http://swoogle.umbc.edu/	0.000022	0.16785	48	0.12429	43
20	http://www.scientificamerican.com/print_ve...	0.000022	0.13367	68	0.09142	66

We computed the equivalent of Figure 5 for the interestingness measure, i. e., we show the rankings for those resources that were within the Top 5 for any of the months. As our time window was one month, we used $C = 1500$ instead of $C = 50$ as in the original paper which used a one-day window. For lack of space and because the diagram did not yield any clear structure, we omit the diagram and summarize the findings.

The top resources were more volatile than in our method. I. e., in our approach, ten different resources made up the top five over all months. In the interestingness computation, there were 70 resources, i.e. each month had a new top five; Table 2 shows the top resource for each month. This indicates that the interestingness is more sensitive to momentary changes in the folksonomy than the FolkRank, and makes it harder to discover long- and medium-term trends. In the top resources, there were few general interest pages such as Slashdot or Flickr. Instead, there were more sites that seemed to be popular at one particular moment in time, but to fade soon afterwards. Figure 6 presents those four resources out of the 70 that overlap with Figure 5. It can be seen that while the interestingness shows some more jitter, the results have the same general direction for both computations.

We conclude that the interestingness, while more scalable and lending itself to a sliding-window visualization as in [6] due to its computational properties, lacks the dampening and generalizing effect of the FolkRank computation, so that it is more useful for short-term observations on particular folksonomy elements.

4   Related Work

There are currently only very few scientific publications about folksonomy-based web collaboration systems. Among the rare exceptions are [6] as discussed above, [8] and [14] who provide good overviews of social bookmarking tools with special emphasis

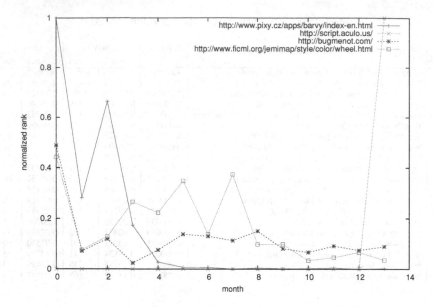

Fig. 6. Evolution of the interestingness values of those resources which overlap with Figure 5; graph plotted the same way as Figure 5

on folksonomies, and [15] who discusses strengths and limitations of folksonomies. The main discussion on folksonomies and related topics is currently only going on mailing lists, e.g. [4]. In [16], Mika defines a model of semantic-social networks for extracting lightweight ontologies from del.icio.us. Besides calculating measures like the clustering coefficient, (local) betweenness centrality or the network constraint on the extracted one-mode network, Mika uses co-occurence techniques for clustering the concept network.

There are several systems working on top of del.icio.us to explore the underlying folksonomy. CollaborativeRank[12] provides ranked search results on top of del.icio.us bookmarks. The ranking takes into account, how early someone bookmarked an URL and how many people followed him or her. Other systems show popular sites (Populicious[13]) or focus on graphical representations (Cloudalicious[14], Grafolicious[15]) of statistics about del.icio.us.

The tool Ontocopi described in [1] performs what is called Ontology Network Analysis for initially populating an organizational memory. Several network analysis methods are applied to an already populated ontology to extract important objects. In particular, a PageRank-like [3] algorithm is used to find communities of practice within individuals represented in the ontology. OntoRank [5] uses a PagesRank-like approach on the RDF graph to rank search results within Swoogle, a search engine for ontologies.

---

[12] http://collabrank.org/

[13] http://populicio.us/

[14] http://cloudalicio.us/

[15] http://www.neuroticweb.com/recursos/del.icio.us-graphs/

Table 2. Top resources for each month according to the interestingness defined in [6]

Month	Resource	Int'ness
0	http://www.pixy.cz/apps/barvy/index-en.html	0.1937
1	http://craphound.com/msftdrm.txt	0.0970
2	http://extensions.roachfiend.com/howto.html	0.1339
3	http://richard.jones.name/google-hacks/gmail-filesystem/gmail-filesystem.html	0.1983
4	http://37signals.com/papers/introtopatterns/	0.2150
5	http://www.fuckthesouth.com/	0.1898
6	http://www.supermemo.com/articles/sleep.htm	0.2585
7	http://www.returnofdesign.com/spectacle/specials/colors.html	0.2958
8	http://www.hertzmann.com/articles/2005/fables/	0.4117
9	http://fontleech.com/	0.4906
10	http://pro.html.it/esempio/nifty/	0.6511
11	http://www.alvit.de/vf/en/essential-...-developers.html	0.5678
12	http://www.newscientist.com/channel/being-human/mg18625011.900	0.6222
13	http://script.aculo.us/	0.8478

Along the same line, in [9], we have presented a technique for analysing ontologies that considers not only the first eigenvector (as PageRank and Ontocopi do), but the full eigensystem of the adjacency matrix of the ontology.

In [2], the evolution of the web graph over time is analyzed. The application of the proposed method lies in the improved detection of current real-life trends in search engines. In comparison to our work, they base their approach on counting timestamped links on pages returned by web searches on given topics, while our contribution infers communities around given users, sites, or topics from the structure of the web graph itself. The algorithm of [2] can currently not be applied to folksonomies, as there exist no folksonomy search engines yet.

Kleinberg [11] summarizes several different approaches to analyze online information streams over time. He distinguishes between three methods to detect trends: using the normalized absolute change, relative change and a probabilistic model. The popularity gradient that we introduced in Section 2.3 is related to the second approach, but differs insofar as it allows for the discovery of *topic-specific* trends, and that we honor steep rises more if they occur higher in the ranking, where the text mining scenario described in [11] requires focusing on words that are neither too frequent nor too infrequent.

5 Conclusion and Outlook

In this paper we have shown how topic-specific trends can be discovered in folksonomy-based systems. The analysis can be done regardless of the types of the underlying resources, which makes folksonomies interesting for multimedia applications.

As folksonomies are still rather young, there are many fascinating research topics left open that are related to the work presented here. They include predicting the change of structure of the folksonomy during its growth, discovering stable and volatile communities, and generating recommendations.

Acknowledgement. This work was partially supported by the European Commission in the projects "NEPOMUK — The Social Semantic Desktop" and "TAGORA - Emergent Semantics in Online Social Communities" under the 6th Framework Programme.

References

1. H. Alani, S. Dasmahapatra, K. O'Hara, and N. Shadbolt. Identifying Communities of Practice through Ontology Network Analysis. *IEEE Intelligent Systems*, 18(2):18–25, 2003.
2. E. Amitay, D. Carmel, M. Herscovici, R. Lempel, and A. Soffer. Trend detection through temporal link analysis. *J. Am. Soc. Inf. Sci. Technol.*, 55(14):1270–1281, 2004.
3. S. Brin and L. Page. The Anatomy of a Large-Scale Hypertextual Web Search Engine. *Computer Networks and ISDN Systems*, 30(1-7):107–117, April 1998.
4. Connotea Mailing List. https://lists.sourceforge.net/lists/listinfo/connotea-discuss.
5. L. Ding, R. Pan, T. W. Finin, A. Joshi, Y. Peng, and P. Kolari. Finding and ranking knowledge on the semantic web. In *International Semantic Web Conference*, pages 156–170, 2005.
6. M. Dubinko, R. Kumar, J. Magnani, J. Novak, P. Raghavan, and A. Tomkins. Visualizing tags over time. In *Proc. 15th Int. WWW Conference*, May 2006.
7. B. Ganter and R. Wille. *Formal Concept Analysis: Mathematical foundations*. Springer, 1999.
8. T. Hammond, T. Hannay, B. Lund, and J. Scott. Social Bookmarking Tools (I): A General Review. *D-Lib Magazine*, 11(4), April 2005.
9. B. Hoser, A. Hotho, R. Jäschke, C. Schmitz, and G. Stumme. Semantic network analysis of ontologies. In Y. Sure and J. Domingue, editors, *The Semantic Web: Research and Applications*, volume 4011 of *LNAI*, pages 514–529, Heidelberg, June 2006. Springer.
10. A. Hotho, R. Jäschke, C. Schmitz, and G. Stumme. Information retrieval in folksonomies: Search and ranking. In Y. Sure and J. Domingue, editors, *The Semantic Web: Research and Applications*, volume 4011 of *LNAI*, pages 411–426, Heidelberg, June 2006. Springer.
11. J. Kleinberg. Temporal dynamics of on-line information streams. In M. Garofalakis, J. Gehrke, and R. Rastogi, editors, *Data Stream Management: Processing High-Speed Data Streams*. Springer, 2006.
12. J. M. Kleinberg. Authoritative sources in a hyperlinked environment. *Journal of the ACM*, 46(5):604–632, 1999.
13. F. Lehmann and R. Wille. A triadic approach to formal concept analysis. In *Conceptual Structures: Applications, Implementation and Theory*, volume 954 of *LNCS*. Springer, 1995.
14. B. Lund, T. Hammond, M. Flack, and T. Hannay. Social Bookmarking Tools (II): A Case Study - Connotea. *D-Lib Magazine*, 11(4), April 2005.
15. A. Mathes. Folksonomies – Cooperative Classification and Communication Through Shared Metadata, December 2004. http://www.adammathes.com/academic/computer-mediated-communication/folksonomies.html.
16. P. Mika. Ontologies Are Us: A Unified Model of Social Networks and Semantics. In Y. Gil, E. Motta, V. R. Benjamins, and M. A. Musen, editors, *ISWC 2005*, volume 3729 of *LNCS*, pages 522–536, Berlin Heidelberg, November 2005. Springer-Verlag.
17. S. Staab, S. Santini, F. Nack, L. Steels, and A. Maedche. Emergent semantics. *Intelligent Systems, IEEE [see also IEEE Expert]*, 17(1):78–86, 2002.
18. L. Steels. The origins of ontologies and communication conventions in multi-agent systems. *Autonomous Agents and Multi-Agent Systems*, 1(2):169–194, October 1998.
19. G. Stumme. A finite state model for on-line analytical processing in triadic contexts. In B. Ganter and R. Godin, editors, *Proc. 3rd Intl. Conf. on Formal Concept Analysis*, volume 3403 of *Lecture Notes in Computer Science*, pages 315–328. Springer, 2005.
20. R. Wille. Restructuring lattice theory: An approach based on hierarchies of concepts. In I. Rival, editor, *Ordered Sets*, pages 445–470. Reidel, Dordrecht-Boston, 1982.
21. W. Xi, B. Zhang, Y. Lu, Z. Chen, S. Yan, H. Zeng, W. Ma, and E. Fox. Link fusion: A unified link analysis framework for multi-type interrelated data objects. In *Proc. 13th International World Wide Web Conference*, New York, 2004.

Personal Semantic Indexation of Images Using Textual Annotations

Grégory Smits, Michel Plu, and Pascal Bellec

France Télécom division R&D,
2, av. Pierre Marzin
F-22307 Lannion Cedex, France
(gregory.smits, michel.plu, pascal.bellec)@francetelecom.com

Abstract. This paper presents an automatic indexation module integrated in online photos management software. Semantic descriptors are generated from textual annotations associated to personal photos. Users naturally annotate their pictures with natural language comments in order to personally describe the main elements of the pictures. Form those personal descriptions, we extract semantic descriptors which are used to organize users' pictures. Our main goal is to retrieve people and places directly or indirectly cited in textual annotations. The descriptors extraction stage is based on a deep linguistic analysis of the textual annotations, which offers a first disambiguation of the possible interpretations and allows for complex descriptors identification (i.e. paraphrases). Paraphrases are then resolved using semantic knowledge sources: a geographical thesaurus and a personal knowledge base of the users' relationships with people. The goal of our system is to automatically integrate new pictures in the user's context accordingly to extracted descriptors. The context that we consider is mainly composed of the current user's taxonomy of descriptors. Thus, our system builds or completes automatically a taxonomy of descriptors which is personalized and relevant for one user.

Keywords: image indexation, semantic indexation, natural language processing.

1 Introduction

We are commonly faced with the management of a huge number of personal pictures. This phenomenon can notably be explained by the prices fall of digital camera, the integration of digital cameras in mobile phones and the possibility to easily share pictures through the Internet. Classifying and organizing personal pictures is a fastidious time consuming task and it might get worse due to the explosion of the number of items each user will produce or receive from others. Thus, an important issue becomes the automation of the indexation task [1] or at least the development of efficient dedicated application to help users organizing their pictures. Someone [2] [3] is an example of such a solution which helps users managing their personal digital documents. Someone solicits users for annotating their pictures with textual

Y. Avrithis et al. (Eds.): SAMT 2006, LNCS 4306, pp. 71–85, 2006.
© Springer-Verlag Berlin Heidelberg 2006

comments and to organize them in a hierarchical and thematic taxonomy of descriptors. This article presents the module we have developed and integrated in Someone to automatically propose some semantic descriptors extracted from the pictures' textual comment. Those proposed descriptors can correspond to user defined descriptors already available in its personal taxonomy, or new ones which have to be correctly inserted in this taxonomy.

In section 2, we first describe the different approaches that can be used to extract relevant descriptors from pictures and their context. Starting from an analysis of a representative corpus of textual annotations of personal photos, we define in Sec. 3.2 the limits of the available indexation methods and consequently the goals of our work. In Sec. 4, the indexation process we have developed is precisely described and illustrated. Section 5 is dedicated to the evaluation of our system. Finally, we enumerate some perspectives and future works in the last section.

2 Related Works

Indexing and retrieving pictures becomes a real issue for millions of people in their everyday life. This is due to the increasing number of pictures available on Internet or on personal storage spaces and the democratization of the use of pictures management software. To deal with this problem, many researches have been already conducted [4]. Different sources of information can be used for pictures indexation. This conducts to commonly consider two alternative approaches [5]: content-based and concept-based approaches.

We will first see in this section that many works are processing information of different natures used to describe pictures content. But we will see in section 3 that some sources of information are not relevant and suitable for personal digital photos software.

2.1 Content-Based and Concept-Based Approaches

Content-based indexing approaches consider the picture as an independent document and analyze graphically the picture independently from its context. Based on image analyzing methods, pictures are described by identified objects, colorimetric measures, size, shapes, type, etc. Despite the fact that methods following this approach directly describe the pictures' components [8] [9], they generally provide low-level descriptors (size, type, colorimetric, color variations, resolution, brightness, etc.). These descriptors are non-intuitive and unnatural. New promising technologies propose some high level features like face recognition (see for example Riya at http://www.riya.com). Bur those sophisticated methods like face or object recognition systems are highly dependant of the picture's quality and the background complexity [4] [10].

For the moment concept-based approaches are the most efficient and successful one [11]. But, this is essentially due to the lack of maturity and the complexity of content-based analyzers. The concept-based approach considers the picture as an item integrated in a more general document. Thus, this method exploits extra-pictural information which constitutes the picture's context. Most popular images search engines like `Google`, `Yahoo` and `picsearch`, are based on this strategy [6]. They principally extract pictures descriptors from: the document title; the URL of the picture; the html ALT tag; the URL of the document and some words surrounding the picture. The main limit of such method is to estimate the relevancy of those information sources. The efficiency of those methods can be easily evaluated on image search engines. If it is relatively acceptable for searching general purpose images on the World Wide Web where a sufficient number of interesting images can be found through millions of images, those methods are not sufficient to retrieve pictures in limited personal collections involving some precise and personal conditions. Such personal conditions can be a picture of a friend or members of your family or being taken at the precise place where you've spend your last holidays.

Considering those two approaches our work is a concept based-approach. Its goal is to improve the level of interpretation and precision of the existing contextual information. This level of interpretation has to be semantic by distinction of a pure lexical interpretation, which means to not consider a descriptor just like a set of characters but a reference to a precise existing entity. The goal of a semantic interpretation is for example to differentiate the several entities which can be identified be a same set of characters, thinking to the town Paris and the surname of a person. One goal is also to identify the entity which can be referred in a textual expression like for example "my brother's girlfriend".

2.2 Indexation in Personal Digital Photos Management Software

Our work is also precisely adapted to the indexation process in pictures management software or online services. In this specific context, indexation is mainly manual. Basically personal pictures management software can be classified in three categories:

- online photo services like `FlickR` (`http://www.flickR.com`) and `Fotoware` (`http://www.fotoware.com`);
- dedicated offline services like `iPhoto`;
- Search functionalities integrated to operating systems (`Spotlight`) on `Mac OS` X and the search assistant on `Windows`).

Both of the previously cited software concerns concept-base approach where the descriptors extraction mainly focuses on extra-pictural sources of information. Indeed, they both take advantages of pictures' titles and user defined meta-information. But pictures titles are generally those which are ascribed by digital camera (img/dsc0001.jpg) and do not give any descriptive information about the document content. That is why users are solicited for commenting their pictures or for associating keywords. Those comments can be expressed as a list of descriptors (keywords) or as a complete natural language annotation. For example, `iPhoto` index and retrieves items regarding their title and associated keywords. But, those keywords are chosen from a limited a priori defined list of possible descriptors

(Family, Children, Holidays, Birthdays, Events, etc.). FlickR, Fotoware and search engines integrated to operating systems do not restrict annotations to limited keywords. Indeed, unlimited keywords and natural language comments can be associated to pictures and used to index them. Users can then retrieve their annotated pictures which are matching a certain query.

Nevertheless photo album software mostly focused on display functionalities and let users managing their personal items manually without any help to index them. This lack can be partly explained by the fact that web pictures indexation methods can not be transposed for personal photos management. Indeed, as we will see in the following section, the particularities of the application context have to be considered in order to propose a relevant and useful automatic indexation system.

3 Requirements of an Automatic Indexation Process for Personal Photo Management

In order to propose an efficient and relevant pictures indexation process, we have first collected a representative corpus of textual annotations associated to personal pictures in photo albums. The goal of this corpus analysis was to identify which descriptors can be extracted from textual annotations to relevantly describe pictures' components and to reuse those descriptors for management tasks (indexation/classification, retrieval).

3.1 Representative Corpus

First of all, we have collected a corpus of textual annotations associated to pictures by their owner from online personal photo albums, personal web sites or blogs. The goal of this preliminary analysis is to validate our subsequent hypothesis of using textual annotations to extract descriptors for pictures indexation. Using queries like "photo album", "personal photo album", "my holidays", "photo gallery", "trip album", etc. submitted to popular web search engines, we have retrieved examples of on-line personal photo albums. This way, we have collected about 30 urls of personal on-line photo albums, which represents 637 pictures annotations. http://www.tripalbum.net is a very good and representative example of photo album web sites used to constitute our corpora.

3.2 Observed Behavior

Our main preoccupation is the development of an automatic indexation process respectful of observed human behaviors. That is why our first step was to identify the nature of the pictures annotations, their meaning and to determine how useful they can be to describe, regroup and retrieve pictures. From a manual analysis of our representative corpus, we have noticed some recurrent behaviors reproduced by people when they annotate their pictures.

First of all, we have noticed the following features. Personal pictures are most often annotated when they are published in a web site (or blog) in order to be shared or publicly displayed. Then, those annotations are mostly individuals, and finally a textual annotation is often associated to a single picture. Concerning the nature of the annotations, they are most of the time short (about 8 words per annotation) and are composed at 90% of a single sentence. We can also notice that annotations are often written in a "good" English or French, respectful of typographic and grammatical rules. We have also observed that annotations mainly focused on recurrent subjects like places, people and events illustrated on pictures.

Of course those observations can be specific to the nature of queries we have submitted to search engines in order to collect our analyzed corpus. Nevertheless, photo album software (on-line and off-line) seems to be used predominantly to manage trip albums and family albums. Thus, about 95% of the annotations we have collected contain a proper name of place or people. Moreover, people and places are not always directly cited in the annotations, but mentioned indirectly using paraphrases[1]. This aspect can notably be explained by pragmatic reasons, like the gathering of pictures in albums or sessions. With on line services, all photos uploaded at the same time are grouped together. This uploading section is sometime identified by a name corresponding to dates, events, holidays, family description, etc. Thus, pictures are annotated according to the group of photos they belong to. Photos softwares also enforce this grouping of photos by folders or albums. Those photos groups partly explained the use of paraphrases, or reference to contextual information in photo annotation in order to avoid repetitions of people and places names and also to recreate continuity between pictures. For example we have found the following kind of annotations of photos in different groups: "We are visiting Italia.", and "And now its capital." in one album and "My mum: Nathalie", "mum is cooking" in another.

The use of paraphrases is an important particularity of annotation of users' pictures which has to be taken in account in order to extract relevant descriptors. Current indexation strategies which consider the pictures textual context as bag of words, only provide inaccurate descriptors for annotations containing paraphrases. Indeed, nominal sentences like "Visit of a famous museum of the French capital." are recurrently used and are frequent in our representative corpus. But, classical pictures indexation methods would only indexed them by informative words: visit, museum, French, capital. Thus, the picture associated to this annotation will not be proposed to users querying photos of Paris.

The previous example also illustrates a major difficulty of this indexation task: ambiguities. And this is particularly the case for people and places proper names (Paris, Nancy, Florence, Adelaide, Albania, etc) which are the main subjects of annotation. This is a reason why it is really important to focus precisely on these two kinds of descriptors. Thus, without a deep analysis of the annotations, only inaccurate and ambiguous descriptors will be extracted which negatively affect the indexation process precision and efficiency.

---

[1] Reformulation of an idea or concept using different words (For example, "The British capital" paraphrase the proper name "London".

3.3 Our Approach Based on a Linguistic Interpretation of the Annotations

In the previous section we have identified some particularities, which should strongly influence the indexation process to develop in order to respect human behaviors.

We have first observed that these pictures and symmetrically their annotations, mostly concern people and places. This explains why Someone and our proposed indexation process essentially focus on these two aspects. Thus, the goal of our process is to extract relevant descriptors of people and places which are then used to index (and symmetrically to retrieve) pictures. But, this information is often ambiguous. For this reason, an adapted disambiguation step is performed in order to identify the correct category of a descriptor (a place or a person). Such ambiguities can be resolved using two kinds of information: the user's context and the linguistic structure of the annotations. Indeed, ambiguous words which can refer to places or people can be disambiguated using user defined contextual information. For example, the extracted descriptor "Paris" which represents an example of such ambiguities should be considered as a first name if the user has manually defined a descriptor of a person named "Paris" or "Pictures of Paris" in the people section of its taxonomy of descriptors managed through Someone. Secondly, the linguistic structure of the textual annotation can also be used to achieve this disambiguation task. In the following annotation "Florence is eating an ice-cream on the beach.", even if "Florence" can refer to a first name or an Italian city, syntactic and semantic aspects should largely influence the indexation process in favor of a first name.

We have also remarked that recurrently people and places are indirectly cited in the annotations using paraphrases. Identifying which entity is referred in those paraphrases is mandatory in order to achieve a precise and relevant pictures indexation. Subsequently, paraphrases have to be identified in order to be resolved. From our corpus, we have noticed that a small set of syntactic patterns can be used to identify such nominal phrases corresponding to paraphrases and that a geographical thesaurus and an ontology of relationship between the user and other people can be used to resolve such linguistic phenomena.

Another issue is when a textual comment refers to a new entity which is not corresponding to an existing descriptor in the user's taxonomy. Then, the indexation process has to propose the creation of new descriptors and their integration in the existing taxonomy.

4 The Indexation Process

This section describes precisely the indexation process which we have developed and integrated in Someone. This process implements the requirements specified in section 3.3. Fig. 1 illustrates the decomposition of the process, which is principally composed of three steps:

1. The candidate descriptors extraction and categorization;
2. Paraphrases resolution;
3. Contextual integration.

To explain the role of each step, we will illustrate each intermediate result generated on the following example which voluntarily representative of the functionalities supported by our system: "My sister Paris is visiting a museum in Florence with Nancy's brother".

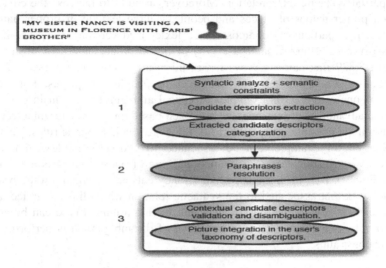

Fig. 1. Decomposition of our indexation process

4.1 Candidate Descriptors Extraction

This first stage aims at extracting candidate descriptors of people and places from textual annotations. Candidate descriptors are composed of places and people proper names or identified paraphrases referring to people and places.

Construction of the linguistic interpretation: In order to identify, extract and categorize candidate descriptors, precise information about the linguistic structure of the annotations is needed. The benefits of indexation methods based on linguistic features have been largely validated [1] [13]. But, due to efficiency reasons they only have been experimented for specific tasks. Nevertheless, as annotations associated to pictures are very short (see 3.2), a linguistic analysis can be performed without damaging the whole performance of the system. Thus, candidate descriptors are extracted from a linguistic interpretation of the annotations. We have identified the linguistic features which can discriminately identify and categorize descriptors representing people and places. Those features mainly concern syntactic relations, lexical and semantic features. For each annotation, a syntactic analysis constrained by semantic rules is achieved by the linguistic toolbox TiLT [14]. Detailed lexical features, semantic features morpho-syntactic categories are affected to each word of the annotation to index. Then, syntactic relations are calculated in order to obtain dependency trees. Fig. 2 graphically illustrates one of the dependency trees proposed by TiLT on our example. Considering the inherent ambiguity of natural languages, several syntactic interpretations can be proposed for a single sentence. The linguistic toolbox has been tuned to perfectly fit our requirements. Indeed, as annotations are not always grammatically well-formed and are mainly composed of a succession of nominal phrases, the grammar has been relaxed in order to generate in every case at least a partial syntactic interpretation. Moreover, in order to increase the coverage of identified proper names of places and people, a large lexicon of proper names has been used. It is particularly composed of a lexicalized version of the Thesaurus Geographical Names data base [15] (475.406 proper names of places) and a lexicon of 8.829 first names. As we have said in a previous section (see 3.2), many proper names are ambiguous and can refer to people and places. For example, "Nancy" is one of the 1.708 examples of lexical ambiguities present in our lexicon. One of the advantages of our indexing approach based on a linguistic interpretation is to constraint the possible interpretations of the words with semantic rules. This means that "Nancy", for example, can be associated to two different lexical units, one qualified by a semantic feature of human and one of place (Nancy is a town in the north of France). But, for a sentence like "Nancy eats an ice-cream on the beach.", a semantic constraint associated to the syntactic relation which links a subject and the verb "to eat", stipulates that only lexical unit having a human type can be attach to this verb with as a subject. Thus, a first level of disambiguation is performed using linguistic constraints.

Candidate descriptors extraction and categorization: Candidate descriptors are extracted according to lexical, syntactic and semantic patterns. Those linguistic features are defined in extraction rules which are written using XSLT and are applied to an XML version of the dependency trees. Each rule (or set of rules) is producing one or several extracted descriptors categorized with one of the five following descriptor types: person item, place item, person paraphrase item, place paraphrase item or a person paraphrase acquisition. Thus each extracted candidate descriptor is directly categorized according to the activated rule. Ambiguities which can't be resolved by the linguistic analysis are left until the contextual integration stage (Sec. 4.3). Tab. 1 illustrates some of the recognition patterns we have developed: Each applied linguistic feature is found in the properties of the nodes in the dependency tree in Fig 2.

Table 1. Linguistic features used to extract candidate descriptors for our base example

Extracted descriptor	Applied linguistic features	Assigned category
"My sister Paris"	lexical: marker of relationship lexical: human proper name	Person paraphrase Acquisition
"Paris"	lexical:1 human proper name	Person
"a museum in Florence"	lexical: place reference lexical: place proper name syntactic: preposition of location	Place paraphrase
"Florence"	Lexical: place proper name	Place
"Nancy's brother"	lexical: human proper name lexical: marker of relationship syntactic: possessive mark	Person paraphrase

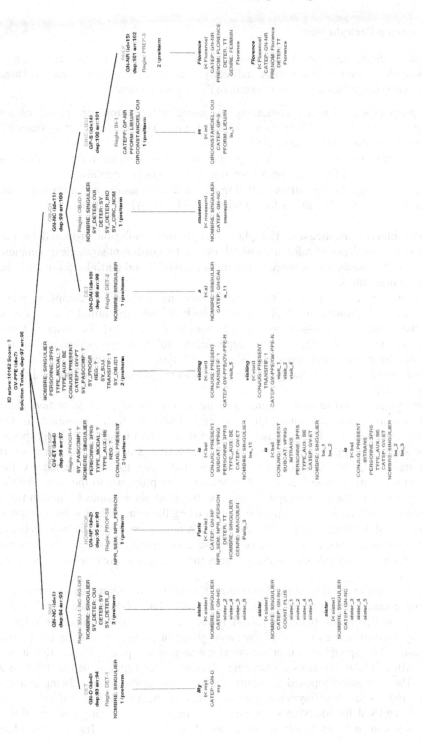

Fig. 2. A graphical representation of a syntactic tree generated by TiLT

4.2 Resolving Paraphrases

Extracted candidate descriptors categorized as paraphrases are then resolved in order to propose the person or place name indirectly cited in pictures' annotations. Three semantic knowledge bases are used to resolve those paraphrases:

1   a geographic thesaurus containing semantic information on place names (e.g. Florence:Lat=43.7830, Long=11.2500, city, Tuscani(region)/Italy(nation)/ Europe (continent) /World(facet)).

2   Ontology of relationships between people including rules for inferring relations from initial ones; like for example the son of a brother is a nephew.

3   A generic linguistic thesaurus including semantic information about lexical units. (e.g. the qualifier "Italian" is linked with the "SEM/ITALY" concept).

Resolving place paraphrases: The place paraphrases resolution proposed by our system covers two types of references which induce two different strategies: complete references ("A museum in Florence", "the Italian capital") and partial references ("the capital", "a museum").

Nominal phrases corresponding to candidate descriptors of complete place paraphrases are translated into SQL queries, which are then submitted to a data base storing the Thesaurus Geographical Names (TGN) [15]. To generate those queries, we take advantage of the dependency sub-tree corresponding to the paraphrase. From such sub-trees, we identify the type of the searched place and additional filters. Basically, the type corresponds to the sub-tree head and additional filters to its children.

Using syntactic and lexical features, sub trees corresponding to identified place paraphrases are translated into SQL queries, which are then submitted to the geographical thesaurus. One can easily imagine that several results will be proposed by the system. When the number of candidate place names is upper than an *a priori* defined threshold, they are marked as unsure and need to be validated by the contextual integration stage (see Sec. 4.3) before being proposed to the user. For the following example "the Italian capital", we search capital place names located in Italy, where Italy is retrieve from the adjective Italian using the generic linguistic thesaurus. Of course, partial references like "the capital" can not be resolved this way, because too many proposals would be generated. We have observed that such references are used to refer to already introduced place names in previous pictures annotations in the same album.

Thus, we use the existing place descriptors in the users' taxonomy of descriptors to search place names which match the partial reference. The TGN is used to validate if a place name descriptor match a partial reference. We also prioritize descriptors previously used to index pictures of the same group of photos. Resolving the partial paraphrase "The capital" in a user's context containing the places names Roma and France, will produce Roma as a matching place as it corresponds in the TGN to a capital. "Paris" is also proposed as it correspond to the capital of an existing country name (country is an hyperonym category of capital in the TGN). Roma is prioritized if it has been used for indexing other photos in the current indexing session. When paraphrases can not be resolved, they are deleted from the list of candidate descriptors.

Resolving references between people: People paraphrases are resolved in order to retrieve people names indirectly cited in textual annotations. This process uses two knowledge sources: a personal file of relationships between the user and other people and an ontology dedicated to personal relationships. Each user of `Someone` is associated to a personal file containing information about its relationships with other people, which can be manually and a priori defined by the user. The content of this file is formatted in the web language `Notation 3`[2], which is then used in order to resolve people paraphrases. The following expressions mean that a person called Paul in the user's relationships has a sister called Nancy: `:Paul :sister : Nancy..`.

The ontology is used to deduce more relationships from initial ones. For example the following rule in this ontology `?A :sister ?B. => ?B :brother ?A.`, can be used to deduce from the previous expressions that Nancy's brother is called Paul. Thus, Paul is proposed as a candidate people descriptor resulting from the resolution of the paraphrase "Nancy's bother" in our example. Each time a new relationship is introduced by the user or acquired from annotation (see below) the deduction task is activated and deduced relationships are stored in the user's personal relationships file.

Identified people paraphrases are translated into Notation 3 queries in order to be computed with the ontology. For example, `:Bart :brother _:WHO.` corresponds to the query used to resolve the paraphrase "Bart's brother". The free variable to guess `_:WHO.` is unified by people names having a relation of `:brother` with `:Bart` according to the current user's personal file of relationships. Unifications of the query variables are then considered as candidate people descriptors for the picture. For compatibility reasons with our application programming language (Java), we have been using the Jena[3] framework to assume this computation. The resolution task and deduction task are performed using the `Euler`[4] inference engine.

Acquisition of interpersonal relationships between people: We have introduced a fifth category of candidate descriptors which corresponds to users' relationships acquisition cases. Indeed, our system distinguishes people paraphrases to be resolved like "Nancy's sister" from acquisition paraphrases "My sister Nancy". Some syntactic patterns have been defined in order to acquire new relationships between the user and other people. Such paraphrases are translated into Notation 3 and then added to the user's personal knowledge source of relationships (if the fact does not already available). For example, "My sister Paris" is translated into `:$USER :sister :Paris` and after having replace $USER by the real user name, the information is added into its semantic knowledge base as a new fact available for future people paraphrases resolution like "Here is my sister", where "Paris" will be proposed as a candidate descriptor corresponding to the resolution of "my sister" (`:$USER : sister _:WHO.`).

---

[2] http://www.w3.org/DesignIssues/Notation3
[3] http://jena.sourceforge.net/tutorial/RDF_API/
[4] www.agfa.com/w3c/euler/

4.3 Contextual Integration, Ambiguity Clarification and Personalization of Proposed Descriptors

Previous modules generate a set of candidate descriptors categorized with one of the two possible facets : people or places. They results from direct extraction based on the linguistic interpretation or from the semantic resolution process of identified paraphrases. Those candidate descriptors are finally integrated in the users' context. The user's context is mainly composed of its personal taxonomy of descriptors and by annotations or indexation proposals laid down by other users on its pictures. Based on this contextual information, candidate descriptors are individually validated as final descriptors if:

-   They match an existing user's descriptor; For example, the candidate descriptor "Nancy" matches the existing user's descriptor "My sisters (Nancy and Sophia)".
-   They share a semantic relation with an existing descriptor; For example, the candidate descriptor "Florence" can be integrated as a sub-descriptor of an existing descriptor "Italia", as these two places are linked in the TGN by a semantic relation of "is-a-city-of". This semantic relation is established using recursively the ancestor field associated to each TGN entry.
-   They match another user proposal; Indexation proposals can be manually laid down by other users and then validate matching candidate descriptors.

For the two first cases previously enumerated, an automatic indexation (i.e. user oriented contextual integration) is performed by our system. Candidate descriptors which can not be directly integrated in the user's context and which are not marked as unsure (see Sec.4.2), are proposed to the user and need to be manually validated by the picture's owner before being created and integrated in the user's taxonomy.

In order to achieve this integration of new descriptors, semantic information is associated to each candidate descriptor during the successive stages of the process (extraction, categorization, paraphrases resolution). For place descriptors, this information gives for example details about the nature descriptor (city, village, country, continent, river, etc.), its ancestors in the geographical thesaurus (like Paris/France/Europe/World). For people candidate descriptors resulting from paraphrases resolution, the stored information is the relationship shared with the user. For example, "Sister/Family" is associated with each result proposed for the resolution of the paraphrase "My sister". When validated by the user and created in the user's taxonomy this contextual semantic information is also kept with the new descriptor. Integration of a new descriptor in the existing taxonomy depends on the semantic information they share. If an existing descriptor corresponds to the semantic information about ancestors associated to a candidate descriptor, then the candidate descriptor is considered as a specialization of the more global existing descriptor (and vice versa). If the new descriptor shares in its semantic information a common ancestor with existing descriptors, they will be group and will have in common another newly created descriptor as ancestor. Thus for example, if "Roma" and "Milan" are existing descriptors in the user's place descriptors, and "Verona" is newly integrated as a validated candidate descriptor for a place, a more general descriptor "Italia" will be also created to regroup these three Italians cities because they share a common ancestor in their contextual semantic information. The same grouping task is achieved for people descriptors when they share a common relationship with the user. For example, the descriptors of two user's brothers can be grouped under a new descriptor named Brothers.

Thus, this last stage of the indexation process can be considered as a personalization of the results, since candidate descriptors are validated using the user's context. Moreover ambiguities which are not resolved during the linguistic analysis ("Florence" can not be linguistically disambiguated in the annotation "Picture of Florence.") can be resolved if contextual information can significantly be in favor of one of the possible categories. For example, if "Italia" is defined as a user's place descriptor then "Florence" will be considered as a place descriptor.

5   An Attempt for Evaluation

To quantify the relevancy of our approach, we have manually evaluated the results proposed by our system on 269 annotations of our base corpus. Classical precision, recall and F-measure metrics [16] have been used to evaluate the candidate descriptors extraction and categorization. Of course, the quality of the linguistic interpretations generated by TiLT strongly influences the relevancy of the whole indexation. Due to the difficulty of evaluating a syntactic parser [17], we have only manually estimated the precision of the generated syntactic interpretation on a small subset of our corpora (50 annotations). We have obtained a precision of about 0.65 and a recall of 0.5. This means that globally, a correct syntactic interpretation (complete or partial) is proposed, but a significant number of erroneous analyses are also generated. The significant presence of noise in the generated set of linguistic interpretations can be explained by the fact that a lot of lexical units are polysemous (15% of the recognized proper names can also be interpreted as nouns, adjective, verb, etc.). Thus, dedicated strategies of word sense and syntactic category disambiguation [18] should be integrated in the parser in order to increase its precision.

Concerning the following stages of the process, we have separately evaluated the relevancy of the extracted candidate descriptors and the relevancy of their categorization:

Table 2. Evaluation of the candidate descriptors extraction and categorization

	Precision	Recall	F-measure
Extraction	0.66	0.8	0.72
Categorization	0.85	0.9	0.87

It is of course impossible to objectively evaluate the efficiency of the paraphrases resolution method and of the contextual validation stage as they principally depend on the user's context state. Thus, we have informally tested the behavior of those stages in order to identify recurrent flaws of our system and possible improvements. Paraphrases and acquisition cases essentially concern nominal phrases, which has to be correctly analyzed by TiLT in order to be efficiently identified. But, considering the heterogeneity of such linguistic phenomenon, this evaluation let us identify which precise and dedicated complementary grammatical rules of nominal phrases recognition have to be developed and notably with relaxed constraints over typographical styles. For example, the following examples should all be considered as the same acquisition case, which is not currently the case: "My sister Nancy", "Nancy, my sister", "My sister (Nancy)", "Nancy: my sister", "My sister - Nancy", etc.

We have also remarked that the contextual validation of candidate descriptors is quite robust and efficient. Ambiguous candidate descriptors like "Paris" in the annotation "Here is Paris!" are well contextually disambiguated when significant information is available. For example, if an entity named Paris is present in the user's Notation 3 file of relationships, then "Paris" as a person name is associated to the picture as final descriptor. Oppositely, if the user's taxonomy of descriptors contains a place descriptor like "Our trip in France", "Paris" is considered as a place descriptor.

6 Conclusions

Personal pictures integrated in digital photo albums are most of the time described by textual annotations. Based on a linguistic analysis of those annotations, the module we have developed extract candidate descriptors corresponding to people and places directly or indirectly cited. Linguistic features are used to identify people and places paraphrases, which are resolved using semantic knowledge sources in order to retrieve the referred person's name or place's name. Finally, our system proposes a contextual validation of the candidate descriptors, which can be considered as a personalization of the indexation process. This process reaches a level of precision in the identification of descriptors which are not currently available in personal picture management services. This free the user in annotating it's picture with natural language comments which are primarily dedicated to communicate and share with other and can be secondly used for retrieval purposes. Moreover, this presented process of semantic descriptor extraction in a textual annotation is also applied to user query in natural language. Users can formulate queries with paraphrases like "photos of my sisters" and it disambiguates queries with proper name being places or person's name. But undeniably, the quality of results of the used syntactic parser affects the whole reliability of the system. This quality depends principally of the linguistic resources which are the lexicon, the grammar rules, and the semantic constrains of those rules. For this reason, we are still increasing the size of our corpus of annotations and conducting evaluation in order to complete linguistic resources for a better coverage and robustness. Anyway, our indexation process achieves promising functionalities which are planed to be integrated in a complete system for semi-automated pictures indexation which will be proposed to user's of an online service for personal pictures sharing.

References

1. Lawrence, S., Giles, C.: Searching the World Wide Web. Science (1998)
2. Agosto, L., Plu, M., Vignollet, L., Bellec, P.: Someone: A cooperative system for personalized information exchange. ICEIS (2003)

3. Plu, M., Bellec, P., Agosto, L., Van De Welde, W.: The web of people: A dual view on the www. Proceedings of The 12th International World Wide Web (2003)
4. Lew , M. Sebe, N. Chabane, D.: Content Based Multimedia Information Retrieval: State of the Art and Challenges, in ACM Transactions on Multimedia Computing, Communications, and Applications, Feb 2006.
5. Azzam, I., Leung, C., Horwood, J.: Implicit concept-based image indexing and retrieval.(2004)
6. Boudry, C., Agostini, C.: Étude comparative des fonctionnalités des moteurs de recherche d'images sur internet. Documentaliste - Sciences de l'information (2004)
7. Smith, J., Chang, S.: An image and video search engine for the world wide web. Proc. IS &T/SPIE Storage & Retrieval for Still Image and Video Databases V, San Jose, (1997)
8. Girgensohn, A., Adcock, J., Wilcox, L.: Leveraging face recognition technology to find and organize photos, ACM Press (2004)
9. Frankel, C., Swain, M., Athitsos, V.: Webseer: An image search engine for the world wide web. (1996)
10. Dori, D., Hal-Or, H.: Semantic content base image retrieval using object process diagram Internation workshop on structural and syntactic pattern recognition (Sydney) (1998)
11. Berinstein, P.: Turning visual: image search engines on the web. Online (1998)
12. Croft, W.: The use of phrases and structured queries in information retrieval. (1991)
13. Fagan, J.: Experiments in Automatic Phrase Indexing for Document Retrieval: A Comparison of Syntactic and non-Syntactic methods. PhD thesis (1987)
14. Guimier De Neef, E., Boualem, M., Chardenon, C., Filoche, P., Vinesse, J.: Natural language processing software tools and linguistic data developed by france telecom r&d. In: Indo European Conference on Multilingual Technologies (IECMT). (2002)
15. The Getty Research Institute, L.A.: Thesaurus geographical names (2006)
16. Larson, R., Learst, M.: Introduction to information retrieval evaluation: Relevance and measures (1998)
17. King, M.: Evaluating natural language processing systems. Commun. ACM (1996) 73–79
18. Ide, N., Veronis, J.: Word sense disambiguation: The state of the art. Computational Linguistics (1998)

Human Activity Language:
Grounding Concepts with a Linguistic Framework

Gutemberg Guerra-Filho and Yiannis Aloimonos

Computer Vision Laboratory,
Department of Computer Science,
University of Maryland,
College Park, MD 20742
guerra@cs.umd.edu, yiannis@cfar.umd.edu

Abstract. We have empirically discovered that the space of human actions has a linguistic framework. This is a sensory-motor space consisting of the evolution of the joint angles of the human body in movement. The space of human activity has its own phonemes, morphemes, and sentences. This has implications for conceptual grounding. We present a Human Activity Language (HAL) for symbolic non-arbitrary representation of visual and motor information. In phonology, we define atomic segments (kinetemes) that are used to compose human activity. We introduce the concept of a kinetological system and propose five basic properties for such a system: compactness, view-invariance, reproducibility, selectivity, and reconstructivity. In morphology, we extend sequential language learning to incorporate associative learning with our parallel learning approach. Parallel learning solves the problem of overgeneralization and is effective in identifying the kinetemes and active joints in a particular action. In syntax, we suggest four lexical categories for our Human Activity Language (noun, verb, adjective, and adverb). These categories are combined into sentences through syntax for human movement.

Keywords: sensory-motor semantic grounding, human activity language, parallel grammatical learning.

1 Introduction

For the cognitive systems of the future to be effective, they need to be able to share with humans a conceptual system. Concepts are the elementary units of reason and linguistic meaning. A commonly held philosophical position is that all concepts are symbolic and abstract and therefore should be implemented outside the sensory-motor system. This way, meaning for a concept amounts to the content of a symbolic expression, a definition of the concept in a logical calculus.

An alternative approach states that concepts are grounded in sensory-motor representations. This sensory-motor intelligence considers sensors and motors in the shaping of the cognitive hidden mechanisms and knowledge incorporation. There exists a variety of studies in many disciplines (neurophysiology, psychophysics,

Y. Avrithis et al. (Eds.): SAMT 2006, LNCS 4306, pp. 86–100, 2006.
© Springer-Verlag Berlin Heidelberg 2006

cognitive linguistics) suggesting that indeed the human sensory-motor system is deeply involved in concept representations.

The functionality of Broca's region in the brain [10] and the mirror neurons theory [4] suggests that perception and action share the same symbolic structure as a knowledge that provides common ground for sensory-motor tasks (e.g. recognition and motor planning) and higher-level activities. Furthermore, spoken language and visible movement use a similar cognitive substrate based on the embodiment of grammatical processing. There is evidence that language is grounded on the motor system [5], what implies the possibility of a linguistic framework for a grounded representation.

In this paper, we investigate the involvement of sensory-motor intelligence in concept description and, more specifically, the structure in the space of human actions. In the sensory-motor intelligence domain, our scope is at the representation level of human activity. We are not mainly concerned with visual perception (motion capture from images), nor with actual motor generation (computation of torque at joints).

We believe multimedia applications will ultimately include all types of sensory data. Current applications involve mostly visual and audio information. Although these media are important, the integration of sensory data with motor information is extremely relevant. An artificial cognitive system with sensory-motor representations is able to learn skills through imitation, better interact with humans, and understand human activities. This understanding includes reasoning and the association of meaning to concrete concepts. The closing of this semantic gap involves the grounding of concepts on the sensory-motor information of the corresponding action. In this paper, we contribute to the grounding of concrete concepts by modeling human actions with a sensory-motor linguistic framework.

The grounding process may start from video, where objects are detected and recognized. At this level, human body parts are features extracted from visual input and, consequently, human movement is captured. In this paper, we are interested in human actions corresponding to general observable voluntary movement.

The problem addressed in this paper is to learn representations for human activity. In this sense, motion capture data is processed towards the discovery of structure in this space. We discovered that human action space has the structure of a language, the Human Activity Language (HAL), with its own phonology, morphology, and syntax. In this paper, we show how we could obtain this language using empirical data. The phonology of human movement involves the segmentation problem, the symbolization problem, and an evaluation system. The morphology of human activity is posed here as a grammatical inference problem.

The availability of a language characterizing human action has implications with regards to the grounding problem, to the universal grammar theory, to the origin of human language and its acquisition process. Besides these theoretical issues, a linguistic representation for human activity has several practical advantages. A compact specification for human activity leads to compression and better efficiency. Once a symbolic linguistic representation is provided, natural language processing and speech recognition are sources of methods that could be applied to activity understanding. A non-arbitrary symbolic representation allows the use of techniques of symbolic reasoning for inference and other cognitive tasks (e.g. recognition) on human activities. This framework could also be used as a basic module of a symbolic query language for the processing of multimedia data.

Human activity representation involves several challenging problems and has many applications in different areas such as Robotics, Kinesiology, Biomechanics, Performing Arts, and Human-Computer Interaction. In Computer Vision, surveillance is achieved with automatic activity detection and recognition based on action representations. They also assist video annotation with efficient storage, transmission, editing, browsing, indexing, and retrieval of the motion data in visual media. Basically, low-level features in the visual data are mapped explicitly or implicitly into higher-level features representing human movement. These features are parsed according to our linguistic framework and, consequently, concrete reasoning is performed on this grounded linguistic space. In Computer Graphics, computer animation performs realistic motion synthesis and composition. A linguistic framework supports automatic animation through the generation of human movement according to the human activity language.

The experimental validation of our linguistic framework is performed in a motion capture database. Our motion capture database contains around 200 different actions corresponding to verbs associated with voluntary observable movement. The actions are not limited to any specific domain. Instead, the database includes actions of several types: manipulative (prehension and dexterity), non-locomotor, locomotor, and interaction. Each action was performed by the same actor repeated times.

The paper follows with a brief review of representative related work in section 2. Section 3 reviews optical motion capture techniques which map motion in visual media to joint angle functions. In section 4, we introduce the concept of kinetology with its five basic properties. The morphology of human movement is described in section 5 through sequential and parallel language learning. In section 6, syntax concerns lexical categories for a Human Activity Language (HAL) and syntactic rules constraining sentence formation. Section 7 summarizes our main results and indicates future research.

2 Related Work

There are several approaches towards bridging the semantic gap between low-level features and high-level concepts. Relevance feedback [11] is an interactive approach for content-based image retrieval. The relevant images are selected according to user feedback and low-level features extracted from each image in a database. Hidden annotation [2] further extends these features by including manually Boolean semantic attributes (e.g. person, city, animal) in the relevance inference. Usually, image databases are only partially annotated due to the heavy manual labor involved. Active learning [15] aims to determine which subset of the database should be annotated. In this sense, our approach is a step towards fully automatic annotation. Given the motion information, each action is automatically converted into our symbolic linguistic representation and linked to the corresponding concept for further processing. Usual text search engines and other symbolic manipulation techniques could be used for the retrieval of multimedia information.

In our linguistic framework, we aim initially to find movement primitives as basic atoms. Fod at al. [3] find primitives by k-means clustering the projection of high-dimensional segment vectors onto a reduced subspace. Kahol at al. [6] use the local minimum in total body force to detect segment boundaries. In Nakazawa at al. [8], similarities of motion segments are measured according to a dynamic programming distance and clustered with a nearest-neighbor algorithm. Wang at al. [13] segment gestures with the local minima of velocity and local maxima of change in direction. The segments are hierarchically clustered into classes using Hidden Markov Models to compute a metric. A lexicon is inferred from the resulting discrete symbol sequence through a language learning approach.

Language learning consists in grammar induction and structure generalization. Current approaches [9, 12, 14] account only for sequential learning. In this paper, we extend language learning to consider parallel learning which is inspired by associative learning.

3   Visual Motion Capture

The mapping from low-level visual features to human movement can be achieved implicitly or explicitly through motion capture. Motion capture is the process of recording real life movement of a subject in some digital geometric representation (e.g. Cartesian coordinates and Euler angles). Optical motion capture uses cameras to reconstruct the body posture of the human performer. One approach employs a set of multiple synchronized cameras to extract markers placed in strategic locations on the body (see Fig. 1).

Fig. 1. Optical motion capture

A more flexible method, markerless monocular (single camera) motion capture (MMMC), avoids the use of markers and extends the capabilities of such systems to any input video. A model-based approach [1] for MMMC uses a 3D articulated model of the human body to estimate the posture such that the projection of the model fits the image of the performer for each frame. Data driven techniques [7] use a motion database to help in the reconstruction of the motion in the video. The motion database is pre-processed in order to create connecting transitions between similar poses according to kinematic features.

Given a video featuring human actions, a MMMC system extracts the human movement from visual features such as silhouettes. Joint angles are computed for all degrees of freedom (DOF) in a hierarchical body model. These joint angle functions are the initial input for segmentation in our linguistic framework.

4   Kinetology: The Phonology of Human Movement

A first process in our linguistic framework is to find structure in human movement through basic units akin to phonemes in spoken language. These atomic units are the building blocks of a phonological system for our Human Activity Language. We refer to this system as a kinetological system. We propose this concept of kinetology, where a kinetological system consists in a geometric representation of 3D movement, a specification of atomic states (segmentation), the association of symbols to segments (symbolization), and satisfies some basic principles. We introduce five principles on which such a system should be based: compactness, view-invariance, reproducibility, selectivity, and reconstructivity.

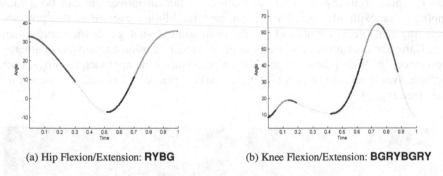

(a) Hip Flexion/Extension: **RYBG**   (b) Knee Flexion/Extension: **BGRYBGRY**

Fig. 2. Symbolic representation of joint angle functions

In order to segment human movement, we consider each actuator independently. An actuator is associated with a joint angle specifying the original 3D motion of the actuator according to an internal geometric representation. The segmentation process assigns one state to each instant of the movement for the actuator in consideration. Contiguous instants assigned to the same state belong to the same segment. We define a state according to the sign of derivatives of a joint angle function. The derivatives used in our segmentation are angular velocity and angular acceleration.

Each segment corresponds to an atom α, where α is a symbol associated with the segment's state. The atomic symbols (**B**, **G**, **R**, **Y**), called kinetemes, are the phonemes of our kinetological system. The symbol **B** is assigned to positive velocity and positive acceleration segments; the symbol **G** is assigned to positive velocity and negative acceleration segments; the symbol **R** is assigned to negative velocity and negative acceleration segments; and the symbol **Y** is assigned to negative velocity and positive acceleration segments (see Fig. 2).

4.1 Compactness and View-Invariance

The compactness principle is related to describing a human activity with the least possible number of atoms in order to decrease complexity, improve efficiency, and allow compression. Compactness is achieved through segmentation which reduces the number of parameters in the representation. Our segmentation approach was implemented as a compression method for motion data and resulted in files with about 3.698% of the original size. Further compression could be achieved with the use of symbolization.

An action representation should be based on primitives robust to variations of the image formation process. A view-invariant representation provides the same 2D projected description of an intrinsically 3D action even when captured from different viewpoints. View-invariance is desired to allow visual perception and motor generation under any geometric configuration in the environment space.

We introduce the Compactness/View-Invariance (CVI) graph for a DOF. A CVI graph shows the states associated with the movement at different viewpoints (see Fig. 3). In order to evaluate the compactness and view-invariance, a circular surrounding configuration of viewpoints is used. For each time instant (horizontal axis) and for each viewpoint in the configuration of viewpoints (vertical axis), the movement state is associated with a representative color.

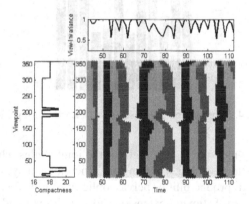

Fig. 3. Compactness/View-Invariance Graph

A compactness measurement consists in the number of segments when the movement varies with time. For each viewpoint, the compactness measurement is plotted on the left side of the CVI graph. The view-invariance measurement concerns the fraction of the most frequent state among all states for all viewpoints at a single instant in time. For each time instant, the view-invariance measure is computed and plotted on the top of the CVI graph.

Note that the view-invariance measure has some uncertainty at the borders of segments and at degenerate viewpoints. The border effect shows that movement segments are not completely stable only during the temporal transition between segments. The degenerate viewpoints are special cases of frontal views where the sides of a joint angle tend to be aligned.

4.2 Reproducibility

An important requirement for a kinetological system is the ability to represent actions exactly in the same way even facing intra-personal or inter-personal variability. A kinetological system is reproducible when the same symbolic representation is associated with the same action performed at different occasions or by different subjects.

A reproducibility measure is computed for each joint angle considering a gait database of 16 people covering males and females at different ages. The reproducibility measure of a joint angle is the fraction of the most frequent symbolic description among all descriptions for the database files. A very high reproducibility measure means that symbolic descriptions match among different gait performances and the kinetological system is reproducible.

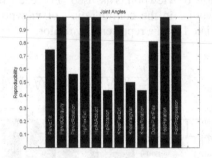

Fig. 4. Reproducibility measure for 12 DOFs during gait

The reproducibility measure is very high for the joint angles which play a primary role in the walking action (see Fig. 4). Using our kinetological system, six joint angles obtained very high reproducibility: pelvic obliquity, hip flexion-extension, hip abduction-adduction, knee flexion-extension, foot rotation, and foot progression. These variables seem to be the most related to the movement of walking forward. Other joint angles obtained only a high reproducibility measure which is interpreted as a secondary role in the action: pelvic tilt and ankle dorsi-plantar flexion. The remaining joint angles had a poor reproducibility rate and seem not to be correlated to the action: pelvic rotation, hip rotation, knee valgus-varus, and knee rotation.

Our kinetological system performance on the reproducibility measure for all the joint angles shows that the system is reproducible for the DOFs intrinsically related to the action. Further, the system is useful in the identification (unsupervised learning) of variables playing primary roles in the activity. The identification of the intrinsic variables of an action is a byproduct of the reproducibility requirement of a kinetological system.

4.3 Selectivity

The selectivity principle concerns the ability to discern between distinct actions. In terms of representation, this principle requires a different structure to represent different actions. We compare the compact representation of several different actions and verify whether their structures are dissimilar.

The selectivity property is demonstrated in our representation using a set of actions performed by the same individual. Four joint angles are considered: left and right hip flexion-extension, left and right knee flexion-extension (see Fig. 5). The different actions are clearly represented by different structures.

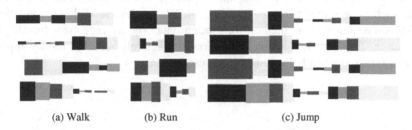

    (a) Walk   (b) Run   (c) Jump

Fig. 5. Selectivity: Different representations for three distinct actions

4.4 Reconstructivity

Reconstructivity is associated with the ability to approximately reconstruct the original movement signal from a compact representation. Once the movement signal is segmented and converted into a non-arbitrary symbolic representation, this compact description is only useful if we are able to recover the original joint angle function or an approximation.

In order to use a sequence of kinetemes for reconstruction, we consider one segment at a time and concentrate on the state transitions between two consecutive segments. Based on transitions, we determine constraints about the derivatives at border points of the segment.

Each segment can have only two possible states for a next neighbor segment. However, the transition **B → Y** (**R → G**) is impossible, since velocity cannot become negative (positive) with positive (negative) acceleration. Therefore, each of the four segment states has only two possible state configurations for previous and next segments and, consequently, there are eight possible state sequences for three consecutive segments. Each possible sequence corresponds to two equations and two inequality constraints associated with first and second derivatives at border points.

A simple model for a joint angle function inside a segment is a polynomial. However, low degree polynomials don't satisfy the constraints originated from the possible sequences of kinetemes. For example, a cubic function has a linear second derivative which is impossible for sequences where the second derivative assumes zero value at the borders and non-zero values at interior points (e.g. **GBG**). The least degree polynomial satisfying the constraints imposed by all possible sequences of kinetemes is a fourth degree polynomial.

The reconstruction process finds five parameters defining this polynomial with the two equations associated with the particular sequence of kinetemes. This involves an under-constrained linear system and, consequently, additional constraints are required to reconstruct the polynomial modeling the joint angle function in a segment. We introduce two more equations using the joint angle value at the two border points. With four equations, a linear system is solved up to one variable. This last free variable is constrained by four inequalities and it can be determined using some criteria such as jerk minimization (see Fig. 6). We implemented this reconstruction scheme as a decompression method for motion data. The average error for all joints was about 0.823°. This way, action generation is effectively achieved just by applying reconstruction to our symbolic compact representation.

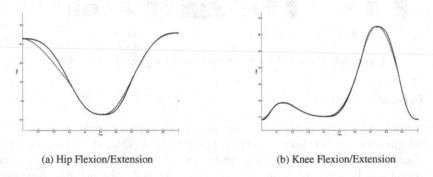

(a) Hip Flexion/Extension   (b) Knee Flexion/Extension

Fig. 6. Reconstruction of joint angle functions

4.5 Symbolization

The kinetological segmentation process results into atoms observing some natural variability. Our goal is to identify the same kineteme amidst this variability. A classification process associates each kineteme with a class/cluster which is denoted as a literal symbol. All kinetemes are obtained by executing the same process for a vocabulary of human actions.

Symbolization consists in clustering motion segments such that each class contains variations of the same motion. This way, each segment is associated with a symbol representing the cluster that contains motion primitives with a similar spatio-temporal structure (see Fig. 7).

Each joint angle is considered independently and the movement corresponding to a specific DOF is segmented into a sequence S of n atoms α_i. The input for our algorithm is this sequence of atoms. A class label is assigned to each atom such that the similar kinetemes will be assigned to the same cluster.

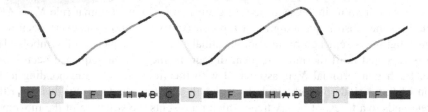

Fig. 7. Segmentation and symbolization

A simple way to perform symbolization is to compute a graph, where the set of vertices corresponds to all segments with the same atomic state. There exists an edge between two vertices in the graph if the similarity distance between the two corresponding segments is less than a threshold value. The similarity distance is the absolute difference between the time normalized versions of the joint angle functions associated with the segments. The symbolization clusters are the connected components of the similarity graph.

5 Morphology

Morphology is concerned with the structure of words, the constituting parts, and how these parts are aggregated. In the context of a Human Activity Language, morphology involves the structure of each action and the organization of a praxicon (lexicon of human movement) in terms of common subparts. Our methodology consists in determining the morphology of each action in a praxicon and then in finding the organization of the praxicon.

The morphology of a specific human action should include the selection of which joints are involved in the activity, the identification of the kinetemes associated with each participating actuator, and the synchronization rules among kinetemes in different active joints. We pose the problem of learning the morphology of a human action as the grammatical inference of a grammar system modeling the human activity such that each component grammar corresponds to an actuator.

5.1 Sequential Grammar Learning

In order to analyze the morphology of a particular action, we are given the symbolic representation for the movement associated with the concatenation of several performances of this action. For each joint angle, we apply a grammar learning algorithm to this representation, which ultimately consists in a single string of symbols instantiating the language to be learned. Each symbol in the string is associated with a time period.

The learning algorithm induces a context-free grammar (CFG) corresponding to the structure of the string representing the movement. The algorithm is based on the frequency of digrams in the sequence. At each step i, a new grammar rule $N_i \rightarrow AB$ is created for the most frequent digram AB of consecutive symbols and every occurrence of this digram is replaced by the non-terminal N_i in the sequence of symbols. This step is repeated until the most frequent digram in the current sequence occurs only once. Each non-terminal N_i is associated with the time period corresponding to the union of the periods of both symbols A and B. The CFG induced by this algorithm corresponds to a forest of binary trees which represents the structure of the movement (see Fig. 8).

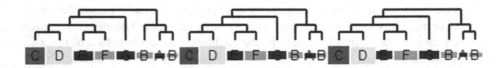

Fig. 8. Grammar forest tree for the hip joint during walk forward

5.2 Parallel Grammar Learning

The execution of a human action involves the achievement of some goal and, therefore, requires consistency in a single string (sequential grammar) and coordination among different strings (parallel grammar). This way, sequential grammar learning and parallel grammar learning are combined to infer the morphology of a human action.

A problem with sequential learning is the overgeneralization that takes place when two unrelated non-terminals are combined in a rule. This happens mostly in higher-levels of the grammar tree, where the digram frequencies are low. In order to overcome this problem, we introduce parallel grammar learning which considers all joint angles simultaneously.

A parallel grammar system consists in a set of simultaneous CFGs related by synchronized rules. A synchronized rule between two non-terminals of different CFGs constrains these non-terminals to have an intersecting time period in the different strings generated by their respective CFGs. This grammar models a system with different strings occurring at the same time. In human activity, each string corresponds to the representation of the movement for one joint angle.

The parallel learning algorithm executes sequential learning considering all the joint angles simultaneously. The digram frequency is still computed within the string corresponding to each joint angle independently. When a new rule is created, the new non-terminal is checked for possible synchronized rules with non-terminals in the CFGs of other joint angles. A synchronized rule relating two non-terminals in different CFGs is issued if there is a one-to-one mapping of their occurrences in the associated strings. Furthermore, any two mapped occurrences must intersect the corresponding time periods (see Fig. 9).

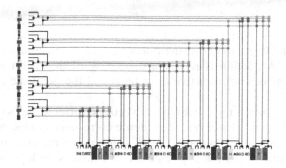

Fig. 9. Parallel grammar system learning

Initially, synchronized rules are difficult to be issued for low-level non-terminals (closer to the leaves of the grammar forest tree). These non-terminals have a high frequency and some atom occurrences are spurious. However, high-level non-terminals are more robust and synchronized rules are reliably created for them. This way, when a synchronized rule is issued for a pair of non-terminals A and B, their descendents in the respective grammar trees are re-checked for synchronized rules. This time, we consider only instances of these descendent non-terminals which are concurrent with A and B, respectively.

We show an execution of our parallel learning algorithm below. For two iterations, we show the set of strings A, the sets of production rules P_i, and the relation R with the synchronized rules. Spaces and dashes are just for visual presentation of the time period associated with each symbol. Non-terminals are displayed only with their index numbers.

```
A = {(a -5- d -5- a d a -5- d -5- a d a -5- d -5- a d),
     (-1- a d a -1- d -1- a d a -1- d -1- a d a -1- d),
     (---4--- ---4--- ---4--- ---4--- ---4--- ---4---),
     (a d a d c a b c a d b b d b c a c d c b b a a d),
     (a -6- d -6- a d a -6- d -6- a d a -6- d -6- a d)},
P1 = {5-> bc}, P2 = {1-> bc}, P3 = {2-> cd, 3-> 2a, 4-> 3b}, P4 = {}, P5 = {6-> bc},
R = {(2,1), (3,1), (4,1), (5,1), (5,2), (5,3), (5,4), (6,1), (6,2), (6,3), (6,4), (6,5)}.

A = {(a -5- d -5- a d a -5- d -5- a d a -5- d -5- a d),
     (-1- a d a -1- d -1- a d a -1- d -1- a d a -1- d),
     (-------7------- -------7------- -------7-------),
     (a d a d c a b c a d b b d b c a c d c b b a a d),
     (a -6- d -6- a d a -6- d -6- a d a -6- d -6- a d)},
P1 = {5-> bc}, P2 = {1-> bc}, P3 = {2-> cd, 3-> 2a, 4-> 3b, 7-> 44}, P4 = {}, P5 = {6-> bc},
R = {(2,1), (3,1), (4,1), (5,1), (5,2), (5,3), (5,4), (6,1), (6,2), (6,3), (6,4), (6,5)}.
```

Besides formally specifying the relations between CFGs, synchronized rules are effective in identifying the maximum level of generalization for an action. Further, the set of joints related by synchronized rules corresponds to the active joint angles concerned intrinsically with the action. The basic idea is to eliminate nodes of the grammar trees with no associated synchronized rules and the resulting trees represent the morphological structure of the action being learned.

6 Syntax

The Subject-Verb-Object (SVO) pattern of syntax is a reflection of the patterns of cause and effect. An action is represented by a word that has the structure of a sentence: the agent is a set of active body parts and the predicate is the motion of those parts.

In a sentence, a noun represents the subjects performing an activity or objects receiving an activity. A noun in a HAL sentence corresponds to the body parts active during the execution of a human activity and to the possible objects involved passively in the action. The initial posture for an action is analogous to an adjective which further describes the active joints representing the noun in a HAL sentence. The sentence verb is associated with the kinematic changes each active joint experiences during the action execution. A HAL adverb models the variation in the execution of each segment. The adverb modifies the verb with the purpose of generalizing the motion. The motion of a segment is represented in a space with a reduced dimensionality, where adverbs are learned.

The lexical categories proposed for HAL compose a nuclear syntax. A HAL sentence S $\rightarrow$ NP VP consists of noun phrase (noun + adjective) and verbal phrase (verb + adverb), where NP $\rightarrow$ N Adj and VP $\rightarrow$ V Adv (see Fig. 10). The organization of human movement is simultaneous and sequential. This way, the nuclear syntax expands to parallel syntax and sequential syntax. The parallel syntax concerns simultaneous activities represented by parallel sentences $S_{t,j}$ and $S_{t,j+1}$. This syntax constrains the respective nouns of the parallel sentences to be different: $N_{t,j} \neq N_{t,j+1}$. This constraint states that simultaneous movement must be performed by different body parts.

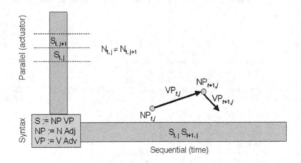

Fig. 10. Nuclear, parallel, and sequential syntax

The temporal sequential combination of action sentences $S_{t,j}$ $S_{t+1,j}$ must obey the cause and effect rule. In sequential syntax, the HAL noun phrase must experience the verb cause and the joint configuration effect must lead to a posture corresponding to the noun phrase of the next sentence. Considering noun phrases as points and verb phrases as vectors in the same space, the cause and effect rule becomes $NP_{t,j} + VP_{t,j} = NP_{t+1,j}$ (see Fig. 10). The cause and effect rule is physically consistent and embeds the ordering concept of syntax.

The lexical units are arranged into sequences to form sentences. A sentence is a sequence of actions that achieve some purpose. In written language, sentences are delimited by punctuation. Analogously, the action language delimits sentences using motionless actions. In general, a conjunctive action is performed between two actions, where a conjunctive action is any preparatory movement that leads to an initial position required by the next sentence.

7 Conclusion

In this paper, we presented a linguistic framework used for conceptual grounding through human activity. The grounding of concepts is achieved through the mapping of concrete verbs (observable voluntary actions) to the sensory-motor representations suggested. We introduced the concept of a kinetological system. We proposed a segmentation approach and a symbolization technique. As an evaluation method, we suggested five basic properties for such a system: compactness, view-invariance, reproducibility, selectivity, and reconstructivity.

In morphology, we extended sequential language learning to incorporate associative learning with our parallel learning approach. Parallel learning solves the problem of overgeneralization and is effective in identifying the active joints, kinetemes, and synchronization (coordination) in a particular action.

In syntax, we suggested four lexical categories for our Human Activity Language (noun, verb, adjective, and adverb). These categories are combined into sentences through nuclear syntax and other specific syntax for human movement: parallel and sequential.

From a methodological viewpoint, our paper introduced a new way of achieving an artificial cognitive system through the study of human action, or to be more precise, through the study of the sensory-motor system. We believe this study represented initial steps of one approach towards conceptual grounding. The closure of this semantic gap will lead to the foundation of concepts into a non-arbitrary meaningful symbolic representation based on sensory-motor intelligence. This representation will serve to the interests of reasoning in higher-level tasks and open the way to more effective techniques with powerful applications.

For future work, we intend to construct a praxicon (vocabulary of actions) and explore its morphological organization towards the discovery of more structure in the human activity language. We also expect more development concerning human movement syntax from the empirical study of this praxicon.

References

1. Chen, Y., Lee, J., Parent, R., and Machiraju, R.: Markerless Monocular Motion Capture using Image Features and Physical Constraints. In Proc. of Computer Graphics International (2005) 36-43.
2. Cox, I., Miller, M., Minka, T., Papathomas, T., and Yianilos, P.: The Bayesian Image Retrieval System, PicHunter: Theory, Implementation, and Psychophysical Experiments. IEEE Transactions on Image Processing, Vol. 9 (2000) 20-37.

3. Fod, A., Mataric, M., and Jenkins, O.: Automated Derivation of Primitives for Movement Classification. Autonomous Robots, Vol. 12, No. 1 (2002) 39-54.
4. Gallese, V., Fadiga, L., Fogassi, L., and Rizzolatti, G.: Action Recognition in the Premotor Cortex. Brain, Vol. 119, No. 2 (1996) 593-609.
5. Glenberg, A. and Kaschak, M.: Grounding Language in Action. Psychonomic Bulletin & Review, Vol. 9, No. 3 (2002) 558-565.
6. Kahol, K., Tripathi, P., and Panchanathan, S.: Automated Gesture Segmentation from Dance Sequences. In Proc. of IEEE International Conference on Automatic Face and Gesture Recognition (2004) 883-888.
7. Lee, J., Chai, J., Reitsma, P., Hodgins, J., and Pollard, N.: Interactive Control of Avatars Animated with Human Motion Data. In Proc. of (2002) 491-500.
8. Nakazawa, A., Nakaoka, S., Ikeuchi, K., and Yokoi, K.: Imitating Human Dance Motions through Motion Structure Analysis. In Proc. of IEEE/RSJ International Conference on Intelligent Robots and Systems (2002) 2539-2544.
9. Nevill-Manning, C., and Witten, I.: Identifying Hierarchical Structure in Sequences: A Linear-Time Algorithm. Journal of Artificial Intelligence Research, Vol. 7 (1997) 67-82.
10. Nishitani, N., Schurmann, M., Amunts, K., and Hari, R.: Broca's Region: From Action to Language. Physiology, Vol. 20 (2005) 60-69.
11. Rui, Y., Huang, T., Ortega, M., and Mehrotra, S.: Relevance Feedback: A Power Tool for Interactive Content-Based Image Retrieval. IEEE Transactions on Circuits and Systems for Video Technology, Vol. 8, No. 5 (1998) 644-655.
12. Solan, Z., Horn, D., Ruppin, E., and Edelman, S.: Unsupervised Learning of Natural Languages. Proceedings of National Academy of Sciences, Vol. 102, No. 33 (2005) 11629-11634.
13. Wang, T.-S., Shum, H.-Y., Xu, Y.-Q., and Zheng, N.-N.: Unsupervised Analysis of Human Gestures. In Proc. of IEEE Pacific Rim Conference on Multimedia (2001) 174-181.
14. Wolff, J.: Learning Syntax and Meanings through Optimization and Distributional Analysis. In Levy, Y., Schlesinger, I., and Braine, M. (Eds.), Categories and Processes in Language Acquisition. Lawrence Erlbaum Associates, Inc., Hillsdale, NJ (1988) 179-215.
15. Zhang, C. and Chen, T.: Active Learning Framework for Content-Based Information Retrieval. IEEE Transactions on Multimedia, Vol. 4, No. 2 (2002) 260-268.

An Engine for Content-Aware On-Line Video Adaptation*

Luis Herranz, Fabricio Tiburzi, and Jesús Bescós

Grupo de Tratamiento de Imágenes, Escuela Politécnica Superior
Universidad Autónoma de Madrid, E-28049 Madrid, Spain
{Luis.Herranz, Fabricio.Tiburzi, J.Bescos}@uam.es

Abstract. In this paper we show a practical implementation of an adaptation engine based on content inspection of video material. The localization of domain independent hints or features that allow to infer the non homogeneous distribution of semantically relevant information, allows to dramatically reduce the amount of adapted data while maintaining the meaningful information. The extraction of these features is performed on-line, via techniques that operate on the compressed domain, following an abstraction model that allows transparent adaptation of DCT based video and wavelets based scalable video.

Keywords: video summarizarion, slideshows, storyboards, video analysis.

1 Introduction

Increasing availability of low cost and high bandwidth connectivity, along with the amazing proliferation of devices with video recording and display capabilities and the subsequent deployment of services in this new scenario are some of the factors that have recently empowered the development of a set of technologies for video content processing and management.

A main topic in this area is the adaptation of audiovisual media to different terminals, networks and users. In this sense, a first challenge is to achieve adaptation (e.g., bitrate reduction, frame size modification, colour conversion) while maintaining most of the *desired* information. A second one is to do this in an efficient way.

Regarding to the first challenge, traditional content-blind adaptation is performed via transcoding: the original media is decoded and then encoded according to the target scenario. Some more efficient solutions, like transcoding without fully decoding[1] or simply extracting bitstream data for scalable formats, via Bitsteam Syntax Description (BSD)[2], are similar from this point of view.

Video coders are currently focused on preserving perceptually relevant information, but do not almost consider semantic relevance; hence, the coding process is usually homogeneous regarding to this concept. However, relevant semantic

---

* Work partially supported by the European Commission under the 6th Framework Program (FP6-001765 - aceMedia Project) and by the Spanish Government under Project TIN2004-07860-C02-01.

Y. Avrithis et al. (Eds.): SAMT 2006, LNCS 4306, pp. 101–112, 2006.
© Springer-Verlag Berlin Heidelberg 2006

information is highly concentrated as spatial or temporal events (i.e., when something *happens*) and shows very low variation elsewhere (i.e., in static scenes, under global and slow motion, or under fast tracking of a relatively static object): it is highly inhomogeneous. Therefore, any adaptation operation that considers semantic features can help reducing the final bitstream size without a significant content loss.

Semantically relevant information is usually highly dependent on the specific domain, ranging from the presence or absence of objects (e.g., people, faces, race cars) to more complex actions (e.g., left objects, people running, overtakes or accidents), which makes it difficult to follow a generic approach. Nevertheless, some meaningful situations can be considered quite domain independent (e.g., shot changes, changes in the camera motion scheme, moving objects relative to the background, etc.), and are hence specially adequate for content-aware domain-independent media adaptation. One of the objectives of the work presented is to extract these semantically relevant features.

Regarding to the second challenge, if we aim to carry out adaptation both over stored and on-line video content, the identification of relevant features has to be on-line performed for every adaptation operation. This avoids the requirement to store semantic hints which may be quite bulky (e.g., segmentation masks), and the need to perform an exhaustive feature extraction which for many adaptation requests would be useless. In order to achieve on-line operation, feature extraction in the presented engine is performed applying techniques that work on the video compressed domain. Some advantages are the direct availability of estimators for features that are hard to extract at a pixel level (e.g., the motion field), and a dramatic reduction of the dimensionality (e.g., by working over DC images); the main drawback is that these techniques are highly dependent on the coding standard and on the coding parameters. Here we present an abstraction model aimed at obtaining features mostly independent of the specific coding domain.

The general objective of this paper is to show a practical implementation of an adaptation engine based on content inspection of coded video material. Section 2 shows a general diagram of the adaptation engine, identifying three modules: Content Analysis, Media Generation, and Adaptation Control. Section 3 presents the analysis module , which integrates the feature extraction techniques into the abstraction model that allows to transparently operate over two notably different domains (DCT based video and wavelets based video). Section 4 introduces the approach followed in the content generation module and the problematic associated to the control of the adaptation process. Finally, Section 5 shows some experimental tests.

2   Overview

This section presents the engine architecture (see Fig. 1) and some design considerations proposed to perform content-driven adaptation of video material. The main difference between this approach and other reported architectures[3][4] for content-aware video adaptation is, first, that we focus on on-line extraction of the semantic features that control the process; and second, that the control loop used to manage the output bit-rate is based in our case on the variation of the number of included relevant frames.

Although we show it here as a standalone component, the adaptation engine presented in this paper is just one of the available content adaptation tools of the framework presented in [5], more specifically the tool devoted to content-aware video adaptation. All the adaptation issues related to media input-output and to the adaptation context are managed by this framework.

The inputs to this engine include the video source that has to be adapted and specific adaptation parameters. Video sources currently include MPEG1/2 video and wavelets based scalable video[6] according to the scheme proposed in [7]. Adaptation parameters include constraints externally imposed by the usage environment: user preferences, network characteristics and terminal capabilities, described via profiles using the MPEG-21 Digital Item Adaptation specification[8]. These are managed by the aforementioned framework, which finally calls this engine specifying a video adaptation mode, frame size requirements, and frame and data rate limits; these are the effective parameters currently accepted by the engine.

Video adaptation modes, which define the available adapted media, include *video slideshows* and *image storyboards*. Frame size modifications are currently performed following a traditional content-blind scheme.

Video slideshows have the same duration as the video source, but include only selected frames in its original temporal location; these frames are then replicated to maintain the original duration. Bandwidth saving is achieved by the small bit-rate employed in the replicated frames, and the semantic relevance is guaranteed by the content-aware analysis performed to select the included frames.

Image storyboards are the result of video transmoding into images. The precise temporal relation between frames and the audio are completely lost but, in compensation, devices that do not support video can have access to some images showing a big picture of what is happening in the video.

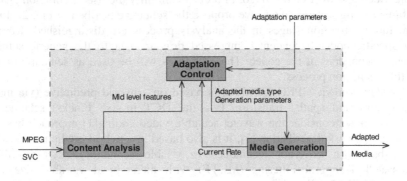

Fig. 1. Overview of the adaptation engine

The three modules identified in the architecture operate as it follows:

– The Content Analysis module is in charge of extracting semantically relevant features from the video sources. The extraction of these features is performed on-line, via techniques that operate on the compressed domain, following an abstraction model that allows transparent adaptation of DCT based video and wavelets based scalable video.

- The Adaptation Control module orchestrates the overall adaptation process. First, according to the adaptation parameters, it commands the Content Analysis module to read the video source and to extract the required features. Then, starting from the extracted features and from an initial estimation of the target bitrate, it performs the selection of the frames that will start conforming the output media. Finally, it launches the Media Generation module establishing a loop that updates the selection of the following frames as a response of the output bitrate.

- The Media Generation module receives a list of frames to start generating the output media in the selected video adaptation mode. In parallel to the generation process, it calculates the output bitrate in order for the Adaptation Control to increase or reduce the number of selected frames.

3 Content Analysis

This section presents a module based on the abstraction analysis model, aimed at obtaining semantically relevant features which are almost independent of the specific coding domain,

As presented in the introduction, the analysis process should be performed on-line, which also requires it to operate in real time. These two conditions impose important constraints on the algorithms' design and on the data used in all the analysis stages. In this sense, compressed domain techniques allow an efficient analysis by avoiding the full decoding of the bitstream and by directly accessing to some compressed domain data useful for analysis.

This module integrates compressed domain analysis in two different codecs: scalable video and MPEG-1/2 video. In order to simplify the use of common analysis algorithms among several codecs we propose the scheme described in Fig. 2. In this design, three extraction stages in the analysis process are distinguished, from the highly specific codec-dependent compressed domain data to the generic semantic features, independent of the codec. These features will be used as semantic hints to guide the adaptation process.

The widely known MPEG-1/2 video is based on a hybrid predictive (via motion compensation) and spatial transform (via the DCT in 8x8 blocks) scheme. For scalable video we consider the wavelet scalable video coding[1] approach based on the t+2D framework described in [7]. It is also based in a hybrid predictive-transform scheme, but with different techniques to enable scalability. It uses Motion Compensated Temporal Filtering (MCTF) with hierarchical variable size block matching (HVSBM) [9] and forward motion compensation. The spatial transform is based on dyadic subband decompositions, similar to those in JPEG2000[10]. It allows spatial, temporal and quality (SNR) scalability. In this paper we will refer to this specific codec as SVC.

Although, for completeness, this section presents an analysis framework devoted to understand the temporal and spatial structure of the video sequence, our work on content-aware adaptation currently considers only temporal features.

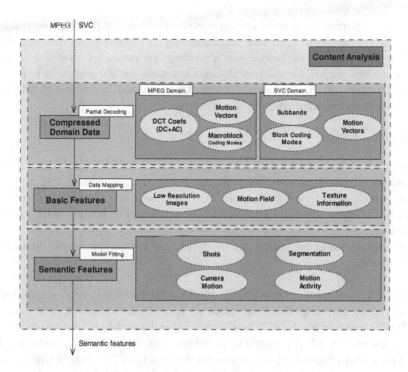

Fig. 2. Architectural overview of the content analysis module

3.1 Compressed-Domain Data

Fast analysis is mainly based on the use of compressed data already present in the bitstream which can be used to infer content features. Only the first stages of the decoding process, usually only entropy decoding, are necessary to extract these required data, avoiding most of the decoding steps and consequently the computational load.

MPEG. The MPEG available compressed domain data are mainly the DCT coefficients (DC and AC) and the macroblock motion vectors. Both can be extracted via header parsing and VLC decoding of the video stream. Due to the heterogeneous structure of the compressed P and B frames, where intra, forward and backward predicted macroblocks can be present in the same frame, the macroblock coding modes should also be considered. Depending on the algorithms and on the requirements in terms of temporal resolution, B and P frames decoding can be avoided (for instance, it's enough to compute a basic feature once per GoP).

SVC. In this case, we can take advantage of the scalability and its structure to extract only the information useful for the analysis. Motion information is present in the header of the bitstream, so that it can be extracted fast without any inverse transform.

Other useful information sources are the subbands which can provide low resolution versions of some frames as well as frequency information at different levels. Basically, the idea is to reduce the decoding as much as possible, avoiding unnecessary inverse MCTF and spatial subband reconstruction stages, only requiring entropy decoding of the subbands. Intrablocks in motion compensated frames are allowed, so block coding modes are also required.

3.2 Basic Features

The purpose of this stage is to get a representation of the video data which is independent of the original coding scheme. The basic elements of this abstraction level are which we refer as "basic features". These will serve as input of the coding independent analysis algorithms devoted to the extraction of features with some kind of semantic relevance.

Three basic features are currently considered: low resolution images, motion field and texture information.

MPEG. DCT coefficients can be exactly obtained for any intra-coded macroblock. In any other case, obtaining the exact value of a coefficient requires in general four block IDCTs and one block DCT; however, this can be avoided if we assume to work with an estimation of the coefficients which can be efficiently obtained [11]. According to this observation, low resolution images (i.e., DC images which are 8x8 times smaller) can be directly estimated from I frames, and quite precisely estimated for B and P frames.

AC coefficients have also proven to be useful to estimate textures and edges[12]. In this sense we use some of these coefficients in two ways: first, to detect blocks containing a single edge in any direction; second, to characterize each block's texture in order to detect texture change. Both indicators are primarily used to reinforce spatial segmentation.

The two previous features are almost coder independent. The motion field can be estimated from the motion estimation vectors, which are highly coding dependent. As a conclusion of our thorough work on this subject[13], we decided to balance between efficiency and vector reliability: we estimate the motion field just from the motion vectors of forward macroblocks in P frames. This yields a sparse motion field (with macroblock resolution).

SVC. Low resolution images are obtained by just selecting the version of the video stream with lower spatial resolution (we consider 3 spatial decompositions, so the low resolution version is 8x8 times smaller than the original). Depending on the frame rate requirements of the algorithms, and on the temporal decompositions available, some inverse MCTF could be necessary.

In MPEG-1/2 the motion field can be obtained directly from the motion vectors present in the bitstream. The SVC implementation that we consider uses variable block size motion compensation: each vector may refer to different block sizes. Therefore, a normalization step is needed to get a motion field referred to a constant block size. This is achieved via replication when the block size is bigger, or merging when smaller, as detailed in [14]. A normalized 16x16 pixels block size is used to have the same reference than for MPEG-1/2.

Texture information can be extracted in a similar way from other subbands, mainly from the lowest temporal subbands. These are the only true temporal low pass subbands, and can be extracted directly from the bitstream without any inverse temporal transform. Thus, these subbands can be considered as an image coded only with spatial dyadic decomposition, where high pass spatial subbands can be used to estimate texture information of the first frame of the GoP at different spatial frequencies.

3.3 Semantic Features

In this section we refer to features more close to a general understanding of what is happening in the video sequence (e.g., the recording camera has changed, the camera begins a zoom, foreground objects move quickly) than to pure signal processing (the motion field, edge indicators, etc.).

These features can not be either obtained directly from the video stream or by a quick mapping of some stream data, so their extraction imposes a great part of the global analysis effort. We currently consider the following semantic features: video shots, motion activity, camera motion and foreground/background segmentation.

Video shots are often used as the basic unit of video temporal segmentation as they group frames with similar content. For video shot detection we use the algorithm proposed in [15] that works directly with frame histogram metrics computed over low resolution images obtained from the previous stage. However as the study in that work was only made for MPEG (where DC images were used) we have carried out similar experiments to evaluate the behavior of this algorithm for different resolution and quality levels. Our results show, as expected, that the performance is comparable for cut detection, while the processing time is much lower, when the lowest spatial resolution version is used.

Motion activity is a visual feature commonly used in tasks as video analysis, content retrieval or video summarization[16] that aims to capture the level of action or pace of a video segment. MPEG-7[17] defines a descriptor to store this feature, computed as the standard deviation of the motion vectors magnitude normalized respect to video rate parameters and later quantized into five levels. We use this descriptor without the quantization step and considering that it was designed for coding schemes with motion vectors referred to a constant block size grid (MPEG-1/2). To compute the motion activity in SVC we use the motion field after the normalization in the previous stage. A comparison between the activities obtained at different temporal levels in SVC and MPEG-1 is described in [14]. In this work, instead of directly using the estimated motion field to obtain this activity descriptor, we first remove camera motion hence achieving an estimation of objects motion.

Camera motion (and especially relevant changes in its scheme) gives useful cues to know where relevant events take place. Likewise the presence of determined camera motions (such as panning or zooms) can be very useful in key frame extraction tasks. To estimate the camera motion the method described in [18] is used, where an iterative algorithm fits the motion vectors to an affine model, rejecting those vectors that don't follow the estimated motion. A coarse segmentation of the foreground moving objects can be obtained from the rejected macroblocks.

4 Adaptation Control and Media Generation

The Adaptation Control is the module that orchestrates the overall operation resulting from a video adaptation request.

As explained in section 2, the adaptation parameters currently include the adaptation mode (either a video slideshow or a storyboard) the target amount of data (the bitrate or the number of images, respectively), and the size and aspect ratio constraints imposed by the terminal.

These two adaptation modes considered so far are based on the selection of semantically relevant frames, that is, on the localization of temporal events. The target amount of data is then considered by the adaptation control in order to decide how many and which relevant frames will finally be included in the adapted media. This process requires to establish some kind of scheme that allows to select as many relevant frames as required (even all, if the case), and some sort of priority policy to guide which specific relevant frames to select.

The diagram in Fig. 3 helps to understand the established mechanisms, which are specifically designed for on-line operation. Let us consider the general case of a video sequence containing several shots. As the video source is being analyzed, the shot change detector marks shot boundaries; these correspond to the first priority level, which is event based. In parallel, the algorithm focused on the detection of changes in the camera motion pattern marks these temporal milestones which correspond to the second priority level, also event based.

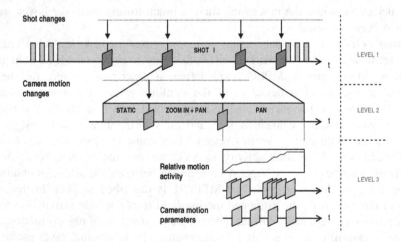

Fig. 3. Frame priority and selection policies in the Adaptation Control

In the absence of events from these two detectors, we can assume that the sequence segment is homogeneous from the camera operation point of view (a same camera and a same motion scheme). The selection of semantically relevant events in this kind of segments begins to be quite arbitrary or domain dependent (e.g., objects entering or leaving, close ups of faces). Hence, in order to keep domain independence, frame selection is here parameter based instead of event based; this corresponds to the third level.

Frame selection in this third level is performed according to two inputs. First, as a result of the algorithm focused on camera motion identification, we account for the motion parameters. The magnitude of these parameters (i.e., the amount of motion) is used to uniformly select frames according to the adaptation control requirements. In parallel, the algorithm devoted to cumulative motion activity estimation, which is applied over camera motion compensated vectors, yields a relative activity curve. An homogeneous subdivision of this curve in the cumulative activity axis (which is on-line achieved when the cumulative value reaches a threshold that can be dynamically modified to select more or less keyframes) is further used to non-uniformly select frames, also according to the requirements.

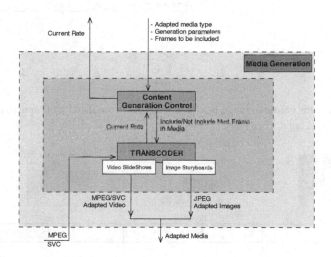

Fig. 4. Architecture overview of the media generation module

The result of the Adaptation Control module is a dynamic table indicating the frames that should be added to the adapted media. The Media Generation module (see Fig. 4) receives this table, along with the remaining required adaptation parameters (mode, spatial constrains, etc.) and commands the transcoder to start encoding while gathering the information about the number of bits used (notice that in some kinds of adaptation even the no-included frames use some bits, as replication has to be made). This number is averaged within a media segment (this segment's size defines the rate accommodation interval) and then sent back to the Adaptation Control module that will update the number and selection of frames that must be included in the adapted media for the following segment.

5 Experimental Tests

In this work we focus on the use of temporal features for domain independent adaptation. A comparative study of the quality of the achieved results could only be achieved via subjective evaluation tests, which have not been carried out so far. The described architecture and the combination of on-line analysis techniques just aim to show the possibility to obtain a much more semantically coherent adaptation that that achieved with content-blind techniques.

In order to illustrate the operation of the proposed engine, it has been tested over a video sequence built merging three well known sequences: akiyo (low activity),

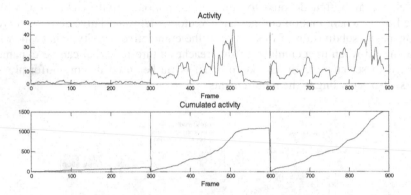

Fig. 5. Shot changes (vertical lines), motion activity and cumulated motion activity of the test sequence

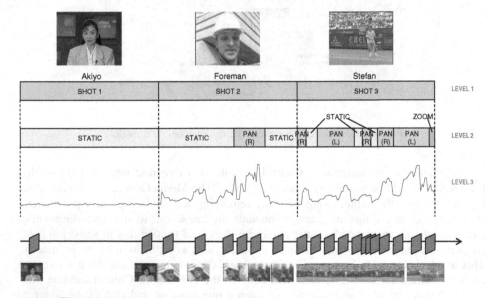

Fig. 6. Sample results of the analysis and adaptation phases in the test sequence

foreman and stefan (medium-high activity, with panning effects). Each one of these sequences has 300 frames with CIF resolution at 30 frames per second. The whole sequence was encoded in MPEG-1, using a GoP of 15 frames and the typical prediction structure IBBPBBPBBB...

Fig. 5 depicts the temporal diagrams obtained for shot changes, motion activity and its cumulated value over a shot. Fig. 6 shows the camera motion scheme throughout the three shots, and an example of frame selection, according to a specific required frame rate, which follows the selection mechanism detailed in the previous section. In this example, frames have been selected synchronized to the occurrence of level one and level two events. In each homogeneous segment defined between every pair of consecutive events, at the third level, the parameters of the camera motion model (e.g. panning velocity) and motion activity have been used to select frames.

6 Conclusions and Future Work

This paper has described an engine aimed at content-aware adaptation of video material. We have focused on the initial ideas that motivate such kind of adaptation, on the reasons that push to perform the adaptation process on-line, on the techniques that allow for on-line extraction of semantically relevant features, and on some mechanisms that finally can infer how many and which frames should be selected in order to generate a meaningful adapted media.

The main conclusion is the feasibility, based on current state of the art analysis techniques, to achieve on-line adaptation of video material taking benefit from the non-homogeneous distribution of the semantically relevant information.

Under the same presented framework, we are currently deepening into the use of generic spatial features (e.g., background-foreground segmentation, region selection according to visual attention, etc.) which may allow both for intra-frame spatial levels to guide variations in each region's quality (similar to the temporal levels described) and for the dynamic selection of the most relevant frame window to achieve content-aware frame size modifications.

References

1. Anthony Vetro, Charilaos Christopoulos, and Huifang Sun, "Video Transcoding Architectures and Techniques: An Overview", IEEE Signal Processing Magazine, vol. 20 (2), Mar. 2003
2. S. Devillers, C. Timmerer, J. Heuer, and H. Hellwagner, "Bitstream Syntax Description-Based Adaptation in Streaming and Constrained Environments", IEEE Transactions on MM, vol. 7(3), Jun. 2005.
3. Vetro, H. Sun, Y. Wang, "Object-Based Transcoding for Adaptable Video Content Delivery", IEEE Transactions on CSVT, Vol. 11, No 3, Mar. 2001.
4. Cavallaro, O. Steiger, T. Ebrahimi, "Semantic Video Análisis for Adapted Content Delivery and Automatic Descriptions", IEEE Transactions on CSVT, Vol. 15, No 10, Oct. 2005.

5. José M. Martínez, Víctor Valdés, Jesús Bescós, Luis Herranz, "Introducing CAIN: a Metadata-driven Content Adaptation Manager Integrating Heterogeneous Content Adaptation Tools", Proc. 6th International Workshop on Image Analysis for Multimedia Interactive Services, WIAMIS 2005, Montreux, Switzerland, Apr. 2005.
6. J.R. Ohm, "Advances in Scalable Video Coding", Proc. of the IEEE, Vol. 93, Issue 1, Jan. 2005.
7. N. Srpljan, M. Mrak, G.C.K. Abhayaratne, E. Izquierdo, "A Scalable Coding Framework For Efficient Video Adaptation", Proc. 6th International Workshop on Image Analysis for Multimedia Interactive Services, WIAMIS 2005, Montreux, Switzerland, Apr. 2005.
8. ISO/IEC 21000-7, Information Technology – Multimedia Frameworks – Part 7: Digital Item Adaptation
9. M. Chan, Y. Yu, and A. Constantinides, "Variable size block matching motion compensation with applications to video coding", Proc. Inst. Elect. Eng., pt. I, Vol. 137, No. 4, pp. 205-212, Aug. 1990.
10. Information Technology – JPEG 2000 Image Coding System – Part 1: Core Coding System, ISO/IEC 15 444-1:2000.
11. Yeo, B. Liu, "Rapid Scene Analysis on Compressed Videos", IEEE Trans. on Circuits and Systems for Video Technology, Vol. 5, No. 6, pp. 533-544, Dec. 1995.
12. M. L. Jamrozik, M. H. Hayes, "A Compressed Domain Video Object Segmentation System", Proc. International Conference on Image Processing, ICIP 2002, Sept. 2002
13. Luis Herranz, Jesús Bescós, "Reliability based optical flow estimation from mpeg compressed data", Proc. International Workshop on Very Low Bit-rate Video, Sardinia, Sep 2005.
14. L. Herranz, F. Tiburzi, J. Bescós, "Extraction of Motion Activity from Scalable-coded Video Sequences", accepted for presentation in SAMT'06, Athens, 2006..
15. J. Bescós, "Real-time Shot Change Detection over On-line MPEG-2 Video", IEEE Transactions on CSVT, 14 (4): 475-484, Apr. 2004
16. Divakaran, R. Radhakrishnan, K.A. Peker, "Motion activity-based extraction of key-frames from video shots", Proc. International Conference on Image Processing, ICIP 2002, (1): 149-152, Sept. 2002
17. ISO/IEC 15938-3: 2002 Information technology. Multimedia content description interface – MPEG-7 Part 3: Visual.
18. V. Mezaris, I. Kompatsiaris, N. Boulgouris, M. Strintzis, "Real-Time Compressed-Domain Spationtemporal Segmentation and Ontologies for Video Indexing and Retrieval", IEEE Transactions on CSV., May 2004

Image Clustering Using Multimodal Keywords

Rajeev Agrawal[1,2], William Grosky[3], and Farshad Fotouhi[2]

[1] Kettering University, 1700 West Third Av.
Flint, MI 48504, USA
[2] Wayne State University, 5143 Cass Avenue,
431 State Hall, Detroit, MI 48202, USA
[3] The University of Michigan – Dearborn,
4901 Evergreen Road, Dearborn, MI 48128, USA
{ragrawal@kettering.edu, wgrosky@umich.edu, fotouhi@wayne.edu}

Abstract. Extending our previous work on visual keywords, we use the concept of template-based visual keywords using MPEG-7 color descriptors. MPEG-7, also called the *Multimedia Content Description Interface*, has been a standard for many years. These color descriptors have the ability to characterize perceptual color similarity and need relatively low complexity operations to extract them, besides being scalable and interoperable. We then demonstrate the power of these visual keywords for image clustering, when used in tandem with textual keyword annotations, in the context of latent semantic analysis, a popular technique in classical information retrieval which has been used to reveal the underlying semantic structure of document collections.

Keywords: MPEG-7, visual keywords, textual keywords, latent semantic analysis, singular value decomposition, adjusted rand index.

1 Introduction

Low-level color and texture image features have been used in the past to classify a large set of images into different clusters [1]. Feature extraction can be based on the entire image or on regions of the image resulting from a segmentation process [2]. Such segmentation techniques have also been used to identify the objects of interest, based on their specific shapes. After segmentation, features are computed from each segmented object and used for clustering. These object segmentation techniques, however, are not very likely to succeed in broad domains [3]. This problem may be circumvented by weak segmentation, where grouping is based on some data-driven properties. Once such features have been extracted, the images are clustered using such methods as k-means, hierarchical agglomerative clustering, or a learning-based approach.

Conventional approaches that use such image attributes as color and texture suffer from a number of problems, such as capturing semantics and formulating queries. One widely popular but highly inefficient solution to this problem is to annotate images with keywords manually, after visually examining them. The image collection can then be queried on these keywords. The quality of this method, however, is

Y. Avrithis et al. (Eds.): SAMT 2006, LNCS 4306, pp. 113–123, 2006.
© Springer-Verlag Berlin Heidelberg 2006

dependent on the perception of the person annotating the images. Even so, this technique is used by many search engines, including Google and Yahoo.

To overcome these problems associated with the above methods, we use both low-level image features, in the form of *visual keywords* [4, 5], and text annotation to cluster the images. These visual keywords result from subdividing an image using templates of certain sizes. As template-based visual keywords are supposed to convey semantics, template sizes are crucial. This is the same problem as in classical information retrieval, where textual keywords are at the word stem level, rather than the individual word level or the paragraph level. Choose a template size too large (akin to a paragraph in classical IR), and it would contain multiple object segments and have a muddled semantics. Choose a template size too small (akin to a letter in classical IR) and its semantics would be completely undetermined.

To organize and search a large text collection, clustering traditionally has been used to discover the inherent concepts embodied there [6]. The document collection can then be organized based on the concepts expressed through these clusters. The basic idea is to extract unique keywords from the set of documents and consider these words as features and then represent each document as a vector of weighted word frequencies in this feature space. A *term-document matrix* is created, in which rows represent the textual keywords and columns represent the documents [7]. Then, techniques such as latent semantic analysis (LSA) can be used. This technique, proposed in [8], uses the truncated singular value decomposition (SVD) to discover the latent relationships between correlated words and documents.

In our approach, we follow the same idea and consider each image as a document and each template region as a word (visual keyword). Hence, each image is represented by multiple template regions. These regions are called *tiles*.

Each tile is represented using the MPEG-7 *scalable color, color structure,* and *color layout* color descriptors. These descriptors have been proven to be very efficient in multimedia content-based search and retrieval [9]. This results, however, in a very large number of distinct tiles. To reduce this large number of tiles, we cluster them and treat tiles in the same cluster as being the same. This is akin to the stemming operator [10] for textual keywords. Thus, in our approach we use a term-image matrix, where the terms consist of textual keywords and visual keywords, each visual keyword being a tile representing a particular cluster of similar tiles. In this paper, we show that this image representation approach produces better image clusters than those resulting from just using image features or just using textual keywords. Although we do not discuss this in detail in this paper, our approach can also be used to associate textual keywords with visual keywords, which can lead to some interesting techniques for annotating images [11].

The rest of the paper is organized as follows. Section 2 provides a brief introduction to the MPEG-7 standard and describes the color descriptors used in this research. In Section 3, we introduce our clustering approach, using both visual keywords represented by MPEG-7 descriptors and already available textual keywords. Experimental results are shown in Section 4. Finally, in Section 5, we discuss future work and offer some conclusions.

2 MPEG-7 Descriptors

MPEG-7, formally called the *Multimedia Content Description Interface*, is a standard for describing multimedia content data that supports some degree of interpretation of semantics determination, which can be passed onto, or accessed by, a device or computer code. MPEG-7 is not aimed at any one application in particular; rather, the elements that MPEG-7 standardizes support as broad a range of applications as possible [12]. MPEG-7 compatible data include still pictures, graphics, 3D models, audio, speech, video, and composition information about how these elements are combined in a multimedia presentation. In this work, the MPEG-7 color descriptors are extracted using a software tool based on MPEG-7 Reference Software: The eXperimentation Model [13]. We expect that a large amount of multimedia data will be represented using the MPEG-7 standard in the coming years. Therefore, it is pertinent to look into the utility of the standard in a variety of applications. There are seven color descriptors: color space, color quantization, dominant colors, scalable color, color layout, color structure, and GoF/GoP color. In its current description, the following six color spaces are supported: monochrome, RGB, YCrCb, HSV, HMMD, and monochrome (intensity only).

What follows is a brief overview of those descriptors which are used in our work:

Scalable color descriptor: The scalable color descriptor (SCD) is a global color histogram, encoded by a Haar transform. The SCD is defined in HSV color space. It has been found to be useful for image-to-image matching and retrieval based on color features. Retrieval accuracy increases with the number of bits used in its representation. The number of bins can be 16, 32, 64, 128 or 256, where, for most applications, it has been found that 64 bits are good enough to use.

Color layout descriptor: The color layout descriptor represents the spatial color information in an image or in an arbitrary shaped region. Being very compact, this descriptor provides for a matching functionality with high retrieval efficiency at very small computational costs. The default number of coefficients is 12.

Color structure descriptor: The color structure descriptor (CSD) captures both color content and the structure of this content. It is used for image-to-image matching and for still image retrieval. An 8×8 structuring element is used to extract color structure information instead of using each pixel individually. This descriptor can distinguish two images in which a given color is present in identical amounts but where the structure of the groups of pixels having that color is different in the two images. The color values are represented in HMMD color space. The number of bins can be 32, 64, 128 or 256. The CSD provides improved similarity-based image retrieval performance compared to ordinary color histograms.

3 Overview of the Proposed Approach

In this section, we discuss our proposed method. The complete process consists of 4 steps:

3.1 Extracting and Clustering Visual Keywords

The most important issue in image search and clustering application is the representation of images using appropriate visual semantics. In [14, 15], tokens have been derived from image pixels which represent the images. The proposed visual keyword idea treats each tile like a word in a textual document. These visual keywords can be used in a variety of applications. Here, we are using them to cluster the images. But, we are confident that they can also be used in annotating, indexing, searching, and retrieving the images.

The algorithm to represent images as visual keywords is given in Figure 1.

Input: A set of images $I = \{I_1, I_2, ..., I_n\}$.
Output: Visual keyword-image matrix
Algorithm:
1. Divide each image I_i into non-overlapping tiles t_i of the fixed template size.
2. Extract MPEG-7 descriptors (SCD, CLD, CSD) to form a tile vector $t_{i,j}$ for each tile t_j of image I_i.
3. Generate a tile matrix V, where each $t_{i,j}$ above is a row vector of V.
4. Normalize V and then apply SVD to reduce the dimension.
5. Apply a clustering algorithm to create C clusters out of all the tiles.
6. Compute the visual keyword-image matrix, having one column for each image and one row for each cluster, where the $(i,j)^{th}$ element of this matrix is the number of times tiles from the ith cluster appear in the jth image.

Fig. 1. Algorithm to create visual keyword matrix

The first step divides the images into tiles using a predefined template size, which do not overlap. **The tiles are selected from left to right and top to bottom manner from each image.** This was a choice we made for our proof-of-concept experiments. The method of selection of tiles will not make any difference to the results. It is like extracting the words from any part of a text document. To avoid complexities, we decided to use non-overlapping tiles of a fixed size, rather than overlapping tiles of many sizes. Based on our successful experiments for these choices, we are currently examining a tile set with these more general properties. In the next step, the MPEG-7 descriptors are extracted from each tile. We decided to use the scalable color descriptor (SCD) with 64 coefficients, which are good enough to provide reasonably good performance [9], the color layout descriptor (CLD) with 12 coefficients, found to be best tradeoff between the storage cost and retrieval efficiency [16], and the color structure descriptor (CSD) with 64 coefficients, sufficient enough to capture the important features of an image [17]. Hence, a tile vector has 140 coefficients. We note that all three MPEG-7 descriptors have different feature spaces; therefore they are normalized within their own feature space using the following simple normalization technique:

$$f_i' = \frac{f_i - \min_i}{\max_i - \min_i}$$

where f_i represents the ith feature in the feature space, $\min_i$ is the minimum value of the ith feature, $\max_i$ is the maximum value of the ith feature, and f_i' is the normalized feature value. The tile matrix is then created using all the normalized tile vectors as its column vectors.

The singular value decomposition (SVD) [18] is then used to reduce the dimension of the normalized tile matrix from 140 to 30. We looked at all 140 eigenvalues, and the first 30 capture all the information, without loss of generality. Using only the first 30 eigenvalues, the resulting matrix approximate the original one with 5 % error [19].

In the next step, a clustering algorithm is applied to cluster the tile matrix into m clusters. We have about 165750 tiles, i.e. 252 tiles per image, generated from the image collection. In our case, **we generate 1500 clusters, which work out to approximately 1% of all the tiles in the collection. We understand that there is no magic formula, which can decide the number of clusters for the tiles.**

We used the *vcluster* algorithm, which is part of the *Cluto* software package [20]. CLUTO is used to cluster low as well as high dimensional datasets and for analyzing the characteristics of the various clusters. We selected this because it is quite fast and open source. The *vcluster* routine uses a method called repeated bisections. In this method, the desired k-way clustering solution is computed by performing a sequence of k − 1 repeated bisections. In this approach, the matrix is first clustered into two groups, and then one of these groups is selected and bisected further. This process continues until the desired number of clusters is found. During each step, the cluster is bisected so that the resulting 2-way clustering solution optimizes a particular clustering criterion function. Note that this approach ensures that the criterion function is locally optimized within each bisection, but in general, it is not globally optimized. The cluster that is selected for further partitioning is the cluster whose bisection will optimize the value of the overall clustering criterion function. The following criterion function is used to find the membership of a tile with a cluster:

$$\text{maximize} \sum_{i=1}^{k} \sqrt{ \sum_{v,u \in s_i} sim(v,u) }$$

The above criterion function is used by many popular vector space variants of the K-means algorithm. In this method, each cluster is represented by its centroid vector and the goal is to find the clustering solution that maximizes the similarity between each vector and the centroid of the cluster to which it is assigned. The similarity between objects is computed using the cosine function, which is basically the dot product of two vectors.

In the final step, an image vector, whose size is the number of clusters, is created for each image. The jth element of this vector is equal to the number of tiles of the given image which belong to the jth cluster. The visual keyword-image matrix is then formed, having all these image vectors for columns. Finally, we normalize each column vector to unit length.

3.2 Creating a Term-Document Matrix Using Textual Keywords

Beyond the use of textual keywords in classical document retrieval [21], text associated with images has been found to be very useful in practice for image retrieval; for example, newspaper archivists index largely on captions [22]. Smeaton and Quigley [23] use Hierarchical Concept Graphs (HCG) derived from Wordnet [24] to estimate the semantic distance between caption words. In this paper, we use textual keywords and combine them with visual keywords to do the clustering. As shown in Section 4 below, the quality of the resulting clusters is much better than what we get using only the textual keywords or only the visual keywords.

The algorithm for this step is very straightforward. We first create an initial term-document matrix (T_{tex}). To control for the morphological variations of words, we use Porter's stemming algorithm [10]. The minimum and maximum term length thresholds are set as 2 and 30, which are reasonable for our experiments. T_{tex} is then normalized to unit-length. In this overall process, we used TMG (Text to Matrix Generator) [25], developed at the High Performance Systems Laboratory at the University of Patras.

3.3 Combining Visual Keywords and Textual Keywords Information

The systems which combine the text with image data include Blobworld [26], where the image segment color is translated into one of a handful of color categories, and image search is then just a simple textual search operation. Webseer [27] uses a similar method to query the images on the web (also see [3]).

In our case, visual keywords are based on MPEG-7 color descriptors derived for each tile cluster of the image and textual keywords are annotations about the image. In order to take the advantage of both visual keywords and textual keywords, both the matrices T_{vis}, T_{tex} are concatenated to create a single large matrix, $T_{vis-tex}$. In our model, the total number of visual keywords is about 3 times larger than the number of textual keywords. Now, we apply LSA on this combined visual and textual space and learn co-occurrence relations among textual keywords and visual keywords. In summary, we extract the semantic relationship between text to text, image to image, and text to image in this step.

3.4 Evaluating the MPEG-7 Visual Keyword Model

Finally, we are interested in clustering the images using the visual + textual keyword model and comparing it with both the visual keyword model and the textual keyword model using the template concept and the template-as-entire-image concept. K-means is one of the simplest unsupervised learning algorithms that can be used to cluster data. This algorithm starts with a number k, which is the desired number of clusters. First k centroids, one for each cluster, are defined, and then each data point is assigned to one of these clusters. This assignment is based on the minimum distance from the data point to the cluster centroids. This procedure can be repeated until centroids do not change their positions.

K-Means minimizes the sum, over all clusters, of the within-cluster sums of point-to-cluster-centroid distances. The squared Euclidean distance measure is used to calculate the distance between data points and centroids.

To compare the estimated class labels after clustering with actual class labels, we have used the adjusted Rand index, described in [28, 29]. The Adjusted Rand Index is a technique for measuring similarity between two data clusters. It has a value between 0 and 1, with 0 indicating that the two data clusters do not agree on any pair of points and 1 indicating that the data clusters are exactly the same.

4 Experiments

In this section, we discuss various experiments using our keyword model. The Corel dataset is very popular to use for any image related experiments, but it has been annotated very carefully to incorporate the correct information about the picture. We wanted to use an image collection which is very diverse and where annotations have some noise in terms of the words used to describe them. Hence, we have used the collection *LabelMe*, available through the MIT AI Lab [30]. This collection allows people to annotate images online and have the annotations be updated instantly. We selected 658 images belonging to 15 categories from this collection. The categories with the number of images in each category are Boston street scene (152), cars parked in the underground garage (39), kitchen (14), office (24), rocks (41), pumpkins (58), apples (11), oranges (18), conference room (28), bedroom (14), dining (63), indoor home (59), home office (19), silverware (81), and speaker (37). Figure-2 has two images from each of the following four categories: office, bedroom, indoor home, and home office starting from the first row and then proceeding left to right, respectively.

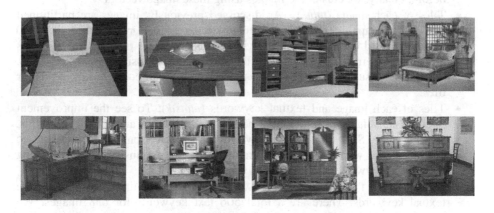

Fig. 2. Images from the categories *office, bedroom, indoor home, home office*

We have used a template size of 32 pixels * 32 pixels to create the non-overlapping tiles for each image. The original images in the collection have different resolutions, which vary from 2560 * 1920 to 300 * 205. The images are resized to 640 pixels * 480 pixels if they are bigger to restrict the number of tiles to a fixed limit; however the smaller images are left in their original sizes. The MILOS software [31], which is based on the MPEG-7 XM model, is used to extract the color descriptors SCD, CSD, CLD. The total number of descriptors used is 140, in which we have 64 of SCD, 64 of

CSD, and 12 of CLD. The maximum number of tiles an image can have is 300; the total number of tiles of 658 images is 165750. We apply a clustering algorithm *vcluster* on these tiles to get 1500 visual keyword clusters. These clusters are then used as visual keywords to create a matrix of 658 images * 1500 clusters.

The textual keyword matrix was created from the annotation list of the image collection and a 658 images * 506 words image-textual keyword matrix is created. The visual keyword and textual keyword matrices are then combined to create a single matrix of size 658 images * 2006 keywords, which contains both types of keywords. LSA/SVD is then applied to select only 200 principal components (coefficients), which results in a matrix of 658 images * 200 concepts.

We have done experiments using the following set of data:

- Full size image (*mpfs*): Clustering is applied on the MPEG-7 color descriptors extracted from the full-size images. Therefore, the visual keyword is actually the entire image and each vector represents the 140 MPEG-7 coefficients. We apply LSA to extract the inherent relationship among images and use only 16 to keep the error rate not more than 5%.
- Full-size image and textual keywords (*mpfstk*): In this case, we also use the textual keywords in addition to visual keywords. After combining both 140 color descriptors and 506 textual keywords, the resultant vector has 646 coefficients for each image. We again apply SVD to extract relationship among images using both types of descriptors and also reduce the dimension of the matrix to 658 images * 38 concepts to maintain the error not more than 5 % as before. Finally, we cluster the images using these image vectors.
- Tiles of each image (*mpts*): As discussed previously, images are partitioned into non-overlapping tiles, and these template-based tiles are clustered into 1500 visual-keyword clusters. After applying SVD, we select only 370 coefficients allowing 5% error. In essence, we cluster the images using visual keywords, where each visual keyword is a tile representing a particular cluster of similar tiles.
- Tiles of each image and textual keywords (*mptstk*): To see the improvement over the *mpts* dataset, we combine both visual keywords and textual keywords. We still apply SVD and surprisingly we need to select only 14 coefficients to allow 5% error with respect to original matrix. Here we cluster the images using both visual keywords and textual keywords.
- In addition to the above datasets, we also cluster the images using only the textual keywords. There are a total 506 text keywords for 658 images. We extract first 142 coefficients using LSA/SVD to keep the same error as in other cases, which sufficiently captures the inherent relationship among the keywords. We find that the results are worse than the result found using any of the above datasets. The adjusted Rand Index is found to be only .26, lowest in all the datasets. This confirms that clustering the images merely on text keywords does not give good results.

The following table shows the result of applying the K-means algorithm for k=15, which is the actual number of categories in the image collection. The adjusted Rand index is used to determine the accuracy of the clusters.

Table 1. Clustering results using different datasets

Dataset	Adjusted Rand Index (ARI)
Mpfs	.32
Mpfstk	.39
Mpts	.34
Mptstk	.51
Text keywords only	.26

The above result shows that when we use MPEG-7 descriptors extracted for full size images (*mpfs*), ARI = 0.32, which is lower than what we get when clustering is done also using textual keywords (*mpfstk*).

The more interesting results are obtained when the images are divided into tiles. We get ARI = 0.34 using only visual keywords (*mpts*), but there is a large improvement we also use textual keywords (*mptstk*). One reason for low values of ARI is that images are not categorized very clearly; there is a lot of overlap in several categories. For example, the categories *office*, *bedroom*, *indoor home*, and *home office*, have similar low-level features. In Figure 3, each row has tiles from the categories shown in Figure 2, but they look very similar to each other. Hence they are likely to have similar low-level features. Based on the type of collection we are using, however, the results are still quite promising. These experiments also show that MPEG-7 descriptors can be used in clustering image collections efficiently. Also, the quality of the clusters can be significantly improved by incorporating the text annotations of the images.

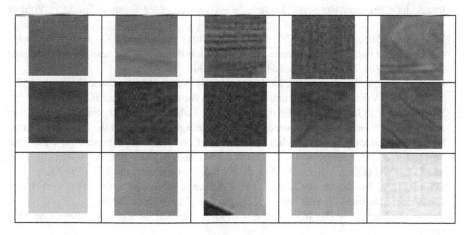

Fig. 3. Tiles from different categories in each row

5 Future Work and Conclusions

This paper presents a new image clustering model using MPEG-7 color descriptors to represent template-based visual keywords, which are then combined with any text keyword annotations. The clustering method also utilizes LSA on the visual keywords and textual keywords of the images. The experiments show that visual keywords and text annotations, if used together, can improve the quality of the clusters. LSA is the key to our approach and helps in establishing the relationship between visual and textual keywords. In [32], salient regions in an image have been used to represent the visual keywords, which are semantically more meaningful, but still involves the complexity of extracting these regions.

We would like to try our model on different datasets, which will help us to compare our results with the results obtained in other researcher's works. There are several possibilities, we would like to try. The text annotations for each image range only from 1 to 10. It is possible to extend this list to include words from other synsets using Wordnet. We have used only three MPEG-7 color descriptors for our experiments. There are other color and texture descriptors, which can be examined. Also, we did not scale the different color descriptors to the same length; obviously the larger size descriptors would have more influence on the initial tile clustering results. Similarly, the number of visual keywords is more than the number of textual keywords. This can be another area of exploration; there is some research needed on how to decide the optimal number of visual and text keywords in an annotated image collection.

References

1. Vailaya, A., Figueiredo, M., Jain, A., Zhang, H.: Image Classification for Content-Based Indexing. IEEE Transaction on Image Processing, Vol. 10, No. 1 (2001)
2. Carson, C., Belonge, S., Greenspan, H., Malik, J.: Blobworld: A System for Region-Based Image Indexing and Retrieval. Lecture Notes in Computer Science. Springer, Volume 1614 (1999)
3. W.M. Smeulders, A., Worring M., Santini, S., Gupta A., Jain, R.: Content based image retrieval at the end of the early years. IEEE Transactions on Pattern Analysis and Machine Intelligence, Vol. 22, No. 12, pp. 1349—1380 (2000)
4. Bhattacharya, A., Ljosa, V., Pan, J., Verardo, M. R., Yang, H., Faloutsos, C., Singh, A. K.: ViVo: Visual Vocabulary Construction for Mining Biomedical Images. ICDM (2005)
5. Sreenath, D.V., Grosky, W. I., Fotouhi, F.: Using Coherent Semantic Subpaths to Derive Emergent Semantics. In Knowledge-Based Intelligent Information and Engineering Systems, Proceedings of the Eighth International Conference (Wellington, New Zealand, September 2004), Lecture Notes in Computer Science, M. Negoita, R. Howlett, L. Jain, and C. Lakhmi (Eds.), Volume 3215, Springer, Berlin, Germany, pp. 173-179 (2004)
6. Dhillon, I. S., Modha, d. S.:Concept Decompositions for Large Sparse Text Data Using Clustering. Machine Learning, vol. 42:1, pages 143-175, January (2001)
7. Salton, G. , McGill, M. J. : Introduction to Modern retrieval, McGraw Hill Book Company (1983)
8. Berry, M. W., Dumais, S. T., O'Brien, G. W. : Using Linear Algebra for Intelligent Information Retrieval. SIAM Review 37(4), 573-595 (1995)
9. Introduction to MPEG-7- Multimedia Content Description Interface. Edited by Manjunath, B. S., Salembier, P., Sikora, T.: John Wiley & Sons, (2002)
10. van Rijsbergen, C.J., Robertson, S.E., Porter, M.F. :New models in probabilistic information retrieval. British Library Research and Development Report, no. 5587, (1980)

11. Barnard, K., Duygulu, P., de Freitas, N., Forsyth, D., Blei, D., Jordan, M. :Matching words and pictures. Journal of Machine Learning Research, 3:1107:1135 (2003)
12. http://www.chiariglione.org/MPEG/standards/mpeg-7/mpeg-7.htm
13. MPEG-7: Visual experimentation model (xm) version 10.0. ISO/IEC/JTC1/SC29/WG11, Doc. N4062 (2001).
14. Turk, M. A., Pentland, A. P.: Eigenfaces for recognition. Journal of Cognitive Neuroscience, 3(1):71-96 (1991)
15. Draper, B. A., Baek, K., Barlett, M. S., Beveridge, J. R.: Recognizing faces with PCA and ICA. Comp. Vis. And Image Understanding, (91):115-137 (2003)
16. Kasutani, E., Yamada, A.: The MPEG-7 Color Layout Descriptor: A Compact Image Feature Description for High-Speed Image/Video Segment Retrieval. ICIP 2001, vol. I, pp. 674-677, October (2001)
17. Manjunath, B. S., Ohm, J. R., Vasudevan, V. V., Yamada, A.: Color and Texture Descriptors. IEEE Transactions on Circuits and Systems for Video Technology, Vol. 11, No. 6 (2001)
18. Deerwester, A., Dumais, S. T., Landauer, T. K., Furnas, G. W., Harshman, R. A.: Indexing by latent semantic analysis. Journal of the American Society of Information Science, 41(6):391-407 (1990)
19. Eckart, C., Young, G.: The Approximation of One Matrix by another of Lower Rank. Psychometrika, 1, pp. 211-218 (1936)
20. CLUTO:A Clustering Toolkit Release 2.1.1, Karypis, G., University of Minnesota, Department of Computer Science, Minneapolis, MN 55455, Technical Report: #02-017, November 28 (2003)
21. Text retrieval Conference. http://trec.nist.gov.
22. Markkula, M., Sormunen, E.: Searching for photos --- journalists' practices in pictorial IR. In the Challenge of Image Retrieval. Electronic Workshops in computing (1988)
23. Smeaton, A. F., Quigley, I.: Experiments on Using Semantic Distances Between Words in Image Caption Retrieval. In Proceedings of SIGIR'96: 174-180 (1996)
24. http://wordnet.princeton.edu/
25. Zeimpekis, D., Gallopoulos, E.: TMG: A MATLAB toolbox for generating term-document matrices from text collections. Technical Report HPCLAB-SCG 1/01-05, Computer Engineering & Informatics Dept., University of Patras, Greece, January (2005) (to appear in Grouping Multidimensional Data:Recent Advances in Clustering, J. Kogan, C. Nicholas and M. Teboulle, eds.,Springer).
26. Carson, C., Belonge, S., Greenspan, H., Malik, J.:Blobworld: Image Segmentation using Expectation-Maximization and its application to image querying. IEEE Transactions on Pattern Analysis and Machine Intelligence 24(8), pp. 1026--38 (2002)
27. Frankel, C., Swain, M. J., Athios, V.: Webseer: An Image Search Engine for the World Wide Web. U. Chicago TR-96-14 (1996)
28. Hubert, L., Arabie, P.: Comparing partitions. Journal of Classification, pages 193–218 (1985)
29. Kuncheva, L. I., Hadjitodorov, S. T.:Using Diversity in Cluster Ensembles. IEEE International Conference on Systems, Man and Cybernetics (2004) pages 1214- 1219 vol.2
30. Russell, B. C., Torralba, A., Murphy, K. P., Freeman, W. T.: LabelMe: a database and web based tool for image annotation. MIT AI Lab Memo AIM-2005-025, (2005)
31. MILOS (http://milos.isti.cnr.it)
32. Tang, J., Hare, J. S. and Lewis, P. H.: Image Auto-annotation using a Statistical Model with Salient Regions (Speech). IEEE International Conference on Multimedia & Expo (ICME 2006)

A Probability-Based Unified 3D Shape Search

Suyu Hou and Karthik Ramani

Purdue Research and Education Center of Information Systems in Engineering (PRECISE)
School of Mechanical Engineering
Purdue University, West Lafayette, U.S.A.
suhou@purdue.edu, ramani@purdue.edu

Abstract. We present a probability-based unified search framework composed of semi-supervised semantic clustering and then a constraint-based shape matching. Given a query, we propose to use an ensemble of classifiers to estimate the likelihood of the query belonging to each category by exploring the strengths from individual classifiers. Three descriptors driven by Multilevel-Detail shape descriptions have been used to generate the classifier independently. A weighted linear combination rule, called MCE (Minimum Classification Error), is adapted to support high-quality downstream application of the unified search. Experiments are conducted to evaluate the proposed framework using the Engineering Shape Benchmark database. The results have shown that search effectiveness is significantly improved by enforcing the probability-based semantic constraints to shape-based similarity retrieval.

1 Introduction

In recent years, 3D shape search has gained importance as a possible means for shape-related engineering knowledge reuse by complementing text-based information systems [1], [2]. Various shape-based descriptors have been developed to support reuse [3]. However, lower level shape representations generated by these techniques do not themselves fully reflect the associated engineering semantics such as function or manufacturing process, thus causing a semantic gap between human understanding and system interpretation of the shape. As a result, the application of knowledge reuse using shape-based retrieval is not very effective.

Various studies [4], [5], [6] have been conducted to reduce the semantic gap in Content-based Search (CBS) by using results obtained from classification. In the engineering domain, it is observed that implicit concurrence relations of geometry cues exist among engineering parts from some semantic categories. However, the situation for classifying an engineering model is different from that of multimedia models/images. There is no unique criterion for classifying engineering models. Even one engineering model sometimes can be classified into different classes by various standards [26]. For example, four parts in Fig.1 have different functions thus belonging to different part families. However they can be classified into the same class by their look. It is hard to use a binary decision when facing with classifying engineering models. Therefore, traditional binary classifier is not generally appropriate for the engineering shape-based classification problem. Hence, it is important to choose a type

Y. Avrithis et al. (Eds.): SAMT 2006, LNCS 4306, pp. 124–137, 2006.
© Springer-Verlag Berlin Heidelberg 2006

of automatic classifier suitable for an engineering context. The classifier ought to be able to predict the classification with high accuracy and with no exclusion to possible choices. Thus, the search engine can prune the database for optimal searching.

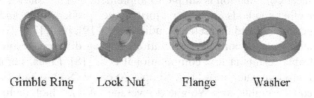

Gimble Ring   Lock Nut   Flange   Washer

Fig. 1. Similar engineering model from different classes

In this paper, a strategy to reduce the semantic gap in our 3D engineering model search system is investigated. The major contribution of this paper is the use of a probability-based classifier in a unified search framework using both semantic concepts and visual content. This paper addresses the problem of nondeterministic classification of engineering models, which can facilitate semantic-constrained shape matching. The interpretation of the probability output is independent of the types of classifiers. We take advantage of this fact to improve the confidence in decision making by classifier combination. The details are elaborated in the following sections. Section 2 introduces the related work. Section 3 outlines the system architecture. Section 4 presents the classifier combination rule which is then employed by the unified search in Section 5. Section 6 discusses the experimental results and concludes in Section 7.

2 Related Works

Recent progress in pattern recognition has made the automatic classification of 3D engineering parts possible by mapping part geometry to engineering semantics. One way of part classification is to recognize the shape pattern embedded in a class by supervised learning. One of the advantages of this way is its simple mathematical interpretation of the pattern from a large set of data extracted from shape. As a result, each class has a general shape pattern encoded in a mathematical configuration with a limited number of parameters. In [9], a weighted k-nearest neighbor KNN is used for engineering part classification. The same research group further applies Support Vector Machines (SVM) to classify the same database and demonstrate that SVM has a better performance than KNN for their classification problem [8]. In [10], active learning is used to employ the information from human labeling to annotate the 3D models automatically. In [11], Bayesian network is used for hierarchical classification. However, the above applications output a binary decision for classification which is not appropriate for the engineering context. In addition, their efforts stop at the classification stage.

Classifier combination has recently been a popular method in various applications of content-based classification [12], [13], [14], [15].Recent studies in combining multiple classifiers for the classification problem has shown proof that the strategy of taking advantage of various resources outperforms traditional monolithic classifiers [16],[18]. Among the various approaches, it is popular to use a linear combination of classifiers that output measurements representing the likelihood of a data belonging to a class [17]. Linear combination is simple to implement and has shown a superior performance over other methods such as majority vote, product rule, and Borda count from many experimental and theoretical studies [18], [19], [20], [21]. Even though the linear combination rules ignore the correlations among different resources, they can still reach plausible results at low computational cost [18]. Linear combination rules can be divided into two categories [17]: Simple Average (SA), which is a linear combination with equal weights; and Weighted Average (WA), which includes rules such as MSE (Minimum Square Error) [22], [23] and MCE (Minimum Classification Error) [24] to estimate the optimal weights for the linear combination model. The main idea is that given a classifier, its contribution to the combined prediction of the testing data is dependent on its performance with the training data. A classifier with better classification accuracy is considered to have better predication capability and will be given more weight for the combination model. However, MSE is criticized for its derivation from regression context rather than for classification considerations [24].

3 System Architecture

In this paper, we use ShapeLab [25] and the Engineering Shape Benchmark (ESB) database [26] as the test bed. ShapeLab has already been well developed for the shape-based search, which provides experimental reference for the proposed research. The classification criteria of ESB are from the methodology developed by Swift and Booker for the purposes of cost estimation and process planning [7], thus reflecting the engineering semantics in this work. Fig. 2 presents the system work flow. First, training data from ESB is used to finalize the individual classifiers and the linear combination model as shown inside the left dotted window of Fig. 2. Each shape descriptor corresponds to a specific classifier. Different classifiers which output the

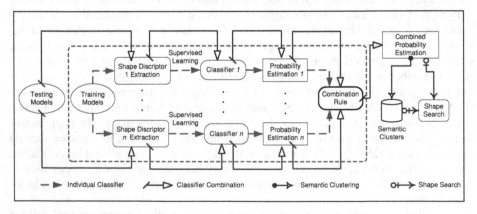

Fig. 2. System architecture on probability-based unified shape search

probability estimation of data being classified to a particular class are developed separately using supervised learning. Meanwhile we exploit the classification output from the training data to estimate the optimal weights of the combination model used later for the real application. The process of how to determine the weights will be discussed in detail in Section 4. After offline initialization, each testing model will go through the shape descriptor extraction and the estimations by individual classifiers before reaching the combination stage. The combined estimation serves two purposes. First, data from the database are deposited into corresponding clusters based on the estimations. We further apply the resulting clusters to the shape matching. Second, the probability-based output determines the share of each cluster for the shape search, or it can allow the user to disambiguate the classification decision by providing a prior list [27]. In either case, the system conducts the shape-based retrieval with preference for some particular clusters of models, thus reducing the semantic gap between the system retrieval and user expectations. In this paper, we mainly focus on the combination rule and its downstream application of the unified search without user interaction.

4 Probability-Based Classifier Combination

The combination rule applied in this paper linearly integrates the individual classifiers which output probability. Before the proposed combination rule is presented, it is necessary to introduce the theoretical background on the probability-based linear combination model.

4.1 Linear Combination Model

Let $x \in \Re^n$ be the input observation vector. The task is to assign x to one of the C possible classes $\Omega = \{\omega_1, \omega_2, ...\omega_C\}$. Define $D = \{D_1, D_2, ..., D_L\}$ to be an ensemble of probability-based classifiers. Each classifier is obtained by some training data. $X = \{(x_i, y_i), i = 1,...N\}$, with $x_i \in \Re^n$ and $y_i \in \Omega$. Let $D_l(x)$ denote the output of the lth classifier D_l for x : $D_l(x) = (d_{l1}(x),...d_{lC}(x))$ where $d_{lj}(x) = P_l(\omega_j \mid x)$ is the likelihood of x being classified to class ω_j by the lth classifier D_l. The estimation satisfies $d_{lj} \in [0,1]$ and $\sum_{j=1}^{C} d_{lj} = 1$.

Define Decision Profile (DP) [28] for input data x as: $DP(x)_{L \times C} = (D_1(x),..., D_L(x))^T$ where T denotes transpose. In this paper, we adopt the linear combination model under Equation (1) for its simple implementation and practical use to achieve plausible results.

$$D_{com}(x;w) = w^T DP(x) \quad \text{where} \sum_{l=1}^{L} w_l = 1 \text{ and } w_l \in (0,1). \tag{1}$$

Let $D_{com}(x;w) = (d_{com,1}(x;w),...,d_{com,C}(x;w))$ as the combined estimation using the linear combination model. $d_{com,j}(x;w) = w^T DP_j(x), j = 1,...C$, where $DP_j(x) = (d_{1j}(x),....d_{Lj}(x))^T$ is the ensemble of the likelihood of data x being assigned to class ω_j from all classifiers. $w = (w_1,...w_L)^T$ is the set of weight factors associated with the L classifiers . The constraint of $\sum_{l=1}^{L} w_l = 1$ with $w_l \in (0,1)$ makes the combined estimation satisfy the condition that $\sum_{j=1}^{C} d_{com,j}(x) = 1$ .Generally, the weight configuration is estimated by some predefined rules using the training output. The commonly referred Simple Average (SA) is the linear combination rule under $w_l = 1/L$. In the next section, the process of how to determine the optimal weights of the linear combination model in Equation (1) is explained.

4.2 Weight Estimation by Adapting MCE

Let $b(x_i) = (b_1(x_i),...,b_C(x_i))$ be a C dimensional class index vector representing the ideal output for the ith data x_i with $b_k(x_i) = 1$ and $b_j(x_i) = 0$ where $x_i \in \omega_k, j \neq k, j = 1,...C,$ and $i = 1,...N$. In this paper, we adopt the $k - fold$ cross-validation to obtain the data for the weight estimation. The cross-validation divides the data into K groups. It uses $K-1$ groups for training and uses the remaining one group to test the classifier obtained from the training data. This procedure repeats K times for each combination of training and testing data. The output of the testing result at each run is collected for the weight estimation

The combination rule in [24] incorporates the MCE criterion for weight estimation. MCE is based on the discriminant function of f in Equation (2) to measure how likely $x \in \omega_k$ is misclassified as another class under the combination rule. In the discriminant function, $f(x;w) \leq 0$ implies a correct decision and $f(x;w) > 0$ indicates a misclassification. The system then obtains the optimal weights by minimizing the overall objective function derived from the discriminant function.

$$f(x;w) = -f_k(x) + \max_{j \neq k} f_j(x) \ , x \in \omega_k .$$ (2)

The MCE rule is targeted to minimize the overall misclassification error from the training data. However, the discriminant function in Equation (2) maximizes the measurement of the right class and minimizes the maximal measurement of the wrong class in a manner of equal importance. The minimization of the misclassification error does not necessarily lead to an increase in the measurement of the right class. In reality, the measurement for the right class is desired to be as big as possible, because it is tightly related to the error between the ideal output and real estimation. Besides, it decides the performance of the unified search for which it is used in this study. Therefore, the minimization of log likelihood error of the probability estimation is also considered for the optimal weight estimation.

$$l(x; w) = \log(b_k(x) - \log(d_{com,k}(x; w)) = -\log(d_{com,k}(x; w))$$

$$\text{where } \log(b_{ik}(x)) = 0 \text{ when } x \in \omega_k.$$

(3)

Unlike Minimum Square Error (MSE) which minimizes the summation of errors from all estimations, the log likelihood error only pays attention to the error corresponding to the right class, thus avoiding the drawback of regression to this context. Obviously the minimization of a log likelihood error has no conflict with the MCE discriminant function. Therefore they can be combined in the same objective function. To unify the target functions, we define the MCE discriminant function in our case as the following:

$$f_i(x) = \log(d_{com,i}(x; w)) \text{ , Therefore}$$

$$f(x; w) = -\log(d_{com,k}(x; w)) + \max_{j \neq k}(\log(d_{com,j}(x; w))) \text{ when } x \in \omega_k$$

(4)

Equation (4) satisfies the same property as Equation (2) that the value of the discriminant function monotonically decreases as the measurement for the right class increases. The final discriminant function to integrate the two desired targets, called Adapted Minimum Classification Error (AMCE) is presented in Equation (5).

$$g(x_i; w) = -\log(d_{com,k}(x_i; w)) + \log(\max_{j \neq k} d_{com,j}(x_i; w)) - \delta\log(d_{com,k}(x_i; w))$$

$$= -(1 + \delta)\log(d_{com,k}(x_i; w)) + \log(\max_{j \neq k} d_{com,j}(x_i; w)) \text{ when } x_i \in \omega_k, i = 1, ...N, \ \delta > 0.$$

(5)

where δ is a user defined parameter to coordinate the preference of minimizing log likelihood error to misclassification error. It is obvious that the discriminant function gives more preference to increasing the measurement of the right class while preserving the ability to minimize the misclassification error.

The sigmoid loss function in Equation (6) is used to smooth the function in Equation (5) to [0,1]. In this paper, ξ is set to 1.

$$l(x_i, w) = \frac{1}{1 + e^{-\xi g(x_i; w)}} \ (\xi > 0).$$

(6)

Finally, the overall loss from all the data is used as the objective function to estimate the weight for the linear combination model. As mentioned earlier in this section, data collected through the cross-validation are used here for weight estimation.

$$\hat{w} = \arg\min_{w} \sum_{i=1}^{N} \sum_{k=1}^{C} l_k(x_i; w) l(x_i \in \omega_k) \text{ with } \sum_{j=1}^{L} \hat{w}_j = 1, \ w_j \in (0,1)$$

(7)

5 Unified Search Framework

5.1 Semantic Clustering

The initial clusters are formed by the ground truth training data, with each cluster associated with a unique engineering semantic meaning. Data from the remaining database is deposited into the corresponding cluster based on the following rule. Let $P_1 = \max_{j=1}^{c}(\{d_{com,j}(x;w)\})$ and $P_2 = \max_{j=1}^{c}(\{d_{com,j}(x;w) \setminus P_1\})$, the clusters corresponding to P_1 and P_2 be labeled C_1 and C_2. If $P_1/P_2 \geq \gamma, (\gamma > 1)$ then $x \in C_1$: otherwise, $x \in C_1$ and $x \in C_2$. The reason to put x into multiple clusters is because it can make up for the false negatives resulting from using only P_1 for clustering. It is possible to deposit x into more than two clusters. However, we chose two for high accuracy of classification without sacrificing to more false positives in the clusters. Through this approach, the geometry models are grouped and indexed at a semantic level where each cluster is associated with a unique semantic meaning.

The proposed approach organizes the data based on the semantic coherency between the data and the patterns extracted from the existing clusters. Hence, by searching within a specific cluster, the problem of semantic gap is fundamentally addressed. The use of the classifier combination enhances classification performance by exploring strengths from different perspectives of shape, thus improving the clustering results. Another advantage of this method is its extensibility. Unlike unsupervised clustering, the structure satiability from the proposed approach is not altered as more and more data are flushed in, as long as the data are still supported by the current classification schema. This clustering approach uses a finite number of labeled data to deal with an infinite number of unlabelled data. In case the unlabelled data do not fit into the current schema, indicated by a large uncertainty measured by entropy, the system will automatically assign a miscellaneous class to it. Thus, the result from the semantic clustering develops an essential foundation for the shape search process which is described in the next section. The combined classifiers can be used repeatedly as long as the overall classification error is still under the designated threshold. The system is designed to update itself from another session of training and validation when the overall classification error crosses the tolerance line.

5.2 Search

The semantic clusters developed in Section 5.1 not only allow the indexing at a semantic level, but also reduce the semantic gap by constraining the shape-based similarity search. The process begins from giving the query the probability estimation for each class using the combined classifiers. The system then performs the shape-based search inside each cluster (Fig.3). The query is compared with the models and the one with the least value is the best match. In this case, the shape search depends on a unified distance which integrates the probability estimation from the combined classifier and shape similarity distance defined in [29] in a single formulation.

$$Unified\ Distance(x, y) = f(Shape\ Similarity(x, y), P(x \in \omega_i \mid x, y \in \omega_i)).\qquad (8)$$

where x is the query and y is the model in the database. $P(x \in \omega_i \mid x, y \in \omega_i)$ is the likelihood of x being in the same class that y belongs to. There are two reasons to choose the probability-based classifier over the binary classifier for the search. First, the inevitable misclassification by binary classifier enlarges the risk of a misled search. This is because the shape similarity search will only retrieve models from the wrong cluster if the query is misclassified by a binary classifier, thus causing a low precision of retrieval. The probability-based classifier mitigates the undesirable consequence from misclassification. Even though the right class does not have the highest probability, it still owns a share of chance. For a good classifier, this value is still going to be higher than most of the incorrect classes. Therefore, the shape search will still retrieve the desired models as a result. By associating the probability with the shape search, it is equivalent to smoothing the zero/one loss function from the binary decision, thus avoiding consequential risk from misclassification. Secondly, the semantic clusters obtained from Section 5.1 are not perfect due to the classification error. Even though the classifier gives the query the highest probability to the correct class, the search is still going to miss some relevant models due to misclassification at the clustering stage. By associating each clustering with a value representing the magnitude of the chance of being the right cluster, the missed models existing in other clusters can still be retrieved. For the same model located in two different clusters, the better unified distance decides its position in the retrieval.

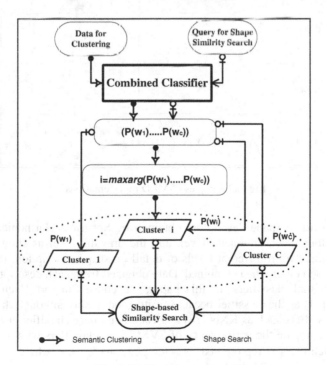

Fig. 3. Unified shape search

In this paper, we examine the retrieval quality of our prototypical design using al the clusters. However, another alternative is to search only those clusters that have likelihood larger than a threshold, or the top t classes where t is a number smaller than the total number of classes. The latter option is better in the case where the number of classes is large. Through this approach, the system can maintain higher precision retrieval at lower computational cost.

6 Experimental Results

6.1 Classifier Selection and Combination

We test the combination rule proposed in Section 4 using real data from the ESB. There are a total of 856 models in ESB. Fifty-five models are miscellaneous and do not belong to any class. The remaining 801 models are grouped into 42 classes. The size of each group varies. The maximum size of a group in ESB is 58, while the minimum size of a group is only 4. Half of the data from each group is randomly selected as the training data. The average training size is 19.6 with a standard deviation of 14.6 which indicates the complexity of our classification problem.

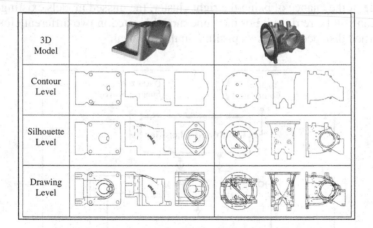

Fig. 4. Examples of MLD representations

We chose three shape descriptors obtained by Spherical Harmonics Transform from the contour level, silhouette level, and the drawing level as shown in Fig. 4. They are obtained from different levels of detail on shape description, thus they can complement each other when combined. Data obtained from each descriptor is used to develop individual classifiers. In this paper, we chose Support Vector Machines (SVM) [30], [31] as the classifier because of its good quality although there are other applicable classifiers such as KNN [28] and Gaussian linear classifier [19]. In this experiment, we compare the performance of SVM-based classifiers with distance-based classifiers: each class is represented by a template of (μ_k, σ_k), where μ_k is the mean and σ_k is the standard deviation of the k_{th} class; the probability output for the classification is obtained by normalizing the distance using Equation (9).

$$P(x \mid \omega_k) = \frac{1/d(x,\omega_k)}{\sum\limits_{i=1}^{C} 1/d(x,\omega_i)}, \text{ where } d(x,\omega_k) = \parallel \frac{x - \mu_k}{\sigma_k} \parallel. \tag{9}$$

Two test cases are conducted to demonstrate the combination performance. In the first case, Case I, the testing data is the remaining half of the database which is not involved in the training. This is the conventional approach to test the classification accuracy by avoiding testing the same data used for training. In the other case, Case II, all the database including the training data are tested for classification accuracy. The reason to perform Case II is because the combined classifier is used to obtain the Precision Recall Curve (PRC) for the unified search. In order to have a fair comparison with PRC from one-shot shape search, we have each example from the database as the query and each model from the database as the target. Table 1 lists the classification accuracy for both test cases. The classifiers include three individual classifiers, the AMCE and the other combination rules such as Simple Average and MCE in Equation (10).

$$g_i(x;w) = -d_{ik}(x;w) + \max_{j \neq k} d_{ij}(x;w)$$

$$l(x,w) = \frac{1}{1 + e^{-\xi g(x;w)}} \quad (\xi > 0) \tag{10}$$

$$\hat{w} = \arg\min_{w} \sum_{i=1}^{N} \sum_{k=1}^{C} l_k(x_i;w) 1(x_i \in \omega_k) \text{ with } \sum_{j=1}^{L} \hat{w}_j = 1, \ w_j \in (0,1).$$

From Table 1, it is apparent that SVM-based classifiers have much better accuracy than distance-based classifiers. The combined estimations outperform each individual classifier, indicating that 1) the system using combined classifiers does not require as much training data as a monolithic classifier in order to reach the same performance; 2) the strategy of classifier combination can certainly help the system to obtain the best performance when no prior knowledge of the individual classifiers is available. The AMCE with $\delta = 2$ is better than other combination rules in all tests.

Table 1. Classification accuracy

	Test Cases	Individual Classifiers				Combination Rules		
		SH_Contour	SH_Silhouette	SH_Drawing	Average	Simple Average	MCE	AMCE(δ=2)
SVM-based	Case I	68.69%	59.71%	59.47%	62.63%	67.23%	68.69%	69.42%
	Case II	72.28%	66.67%	66.29%	68.41%	72.66%	73.78%	75.16%
Distance-based	Case I	38.24%	33.71%	33.62%	30.57%	42.14%	42.09%	42.97%
	Case II	41.32%	35.45%	35.21%	37.33%	46.56%	46.44%	47.69%

Fig. 5 shows the distribution of the number of correct classifications out of 412 testing data within the rank of top 10 by several classifiers. It is apparent that there is only a trivial improvement of classification accuracy after the top 3, which implies that most of the queries will have the measurement of the right class located within the top 3. The relaxed classification accuracy for the top 2 is 84.78% for the proposed method. It indicates that 69.42%-84.78% of the testing data can be correctly deposited into the right cluster based on the parameter γ of the rule in Section 5.1. In this experiment, we chose a higher value of γ to exclude putting the same data into multiple clusters. The main reason is because we want the PRC, an indication of the retrieval quality, to be solely dependent on the probability estimation of the query and the shape similarity distance between the query and the target model. If the same model appears twice in the matching process it will affect the PRC. Therefore it is inappropriate to compare the unified search with the one-shot search method. However, it is plausible to put data into multiple clusters in reality because it will certainly improve the retrieval precision, although at a little more computational cost.

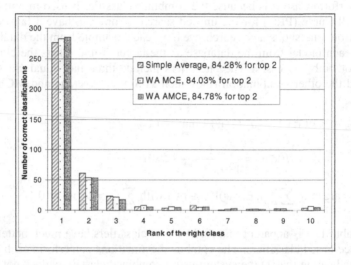

Fig. 5. Number of correct classifications and relaxed classification accuracy for top 2

6.2 Precision and Recall

The Precision Recall Curve (PRC) is used to characterize the performance of the unified search. The unified distance defined in Equation (8) is employed here to compute the similarity between the query and the models. Each of the 801 models is involved in generating the PRC. Finally, the precision values are averaged at a series of uniform recall intervals. The PRC for each individual model is dependent on several factors: i) the probability-based semantic consistency between the query and the models, ii) the results from the clustering, and iii) the shape-based matching. The probability-based semantic consistency means the likelihood of the query and the model belonging to the same category. This consistency is the most critical factor for determining the trend of the PRC. If the right class gets bad estimation, the desired models will then be downgraded in the retrieval list, thus causing lower precision than normal. The system sorts the similar models based on the value of the unified distance throughout all the models. Once all relevant models are retrieved, the system iterates to the next query until all the models have been exhaustively visited. If the classifier gives the correct class the highest probability value, the models inside the right cluster will have the highest priority in the search. Even if the right class does not have the highest estimation, the models inside the right group will not be ignored from the top retrievals if the estimation is still higher that of most of the other classes. As it is indicated in Fig.5, the right class usually has one of the top three likelihoods in the probability estimation. The worst case happens when the classifier gives the lowest measurement to the right class. This is why the classifier combination strategy should be considered in the search because the complementary nature will embrace the strength of each classifier, thus avoiding the worst case.

Fig. 6 shows the PRC of various cases. The solid lines represent the results from the probability-based unified search using SVM-based classifiers, including three individual classifiers and three combined classifiers such as SA, MCE and AMCE. The dotted line represents the PRC using the same configuration however with the distance-based classifiers instead of SVM. The dashed line represents the PRC from the MLD representation which showed a good performance for ESB [29]. Apparently, the PRC from the probability-based search with SVM classifiers are significantly better than the PRC from other methods, which supports the difference of the classification accuracies from Table 1. The high value at the end of the PRC implies that the system can obtain all relevant models far before the database has been exhaustively visited. The PRC of the combined classifiers outperform those of individual classifiers for both SVM-based classifiers and distance-based classifiers. Evidently, the PRC performance has a direct relationship to the overall classification error. The PRC of SH from the silhouette level and the PRC of SH from the drawing level behave similarly because their classification accuracies are close. The AMCE improves over the individual classifiers and has a slightly better performance than other selected combination rules. In order to evaluate the performance of the unified search compared to the one-shot search method, we adopt the Average of Difference (AOD) employed in[26] to quantify the difference. We calculate the average of the differences between

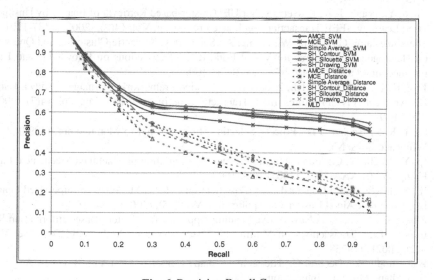

Fig. 6. Precision Recall Curves

the precisions value of the MLD and those of AMCE at all recall intervals. Our results shows that, on average, the proposed search based on AMCE can achieve 23.2% higher precision than that of the MLD alone.

7 Conclusions

In this paper, we presented a framework to tackle the semantic gap of shape-based search using a probability-based classifier. We adapted MCE by integrating the targets of minimizing the classification error and the log likelihood error into a single formulation. The adapted combination rule is further used to enhance the classification accuracy for the good of the unified search. Experiments using our ESB demonstrate that the proposed combination rule is better than individual classifiers. The accuracy improvement is 6.8% for case I and 6.72% for case II over the average performance of the individual classifiers. The relaxed classification accuracy can reach about 85% for the top 2 and 90% for the top 3 in case I. The PRC obtained by the proposed search has superior performance supported by the fact that the AOD of the proposed PRC is 23.2% higher than the PRC obtained from MLD alone.

References

1. Cardone, A., Gupta, S.K., Karnik, M.: A Survey of Shape Similarity Assessment Algorithms for Product Design and Manufacturing Applications, ASME Journal of Computing and Information Science in Engineering, Vol. 3(2), (2003) 109-118
2. Iyer, N., Lou, K., Jayanti, S., Kalyanaraman, Y., Ramani, K.: Shape-Based Searching for Product Lifecycle Applications, Computer Aided Design, Vol. 37(13), (2005) 1435-1446
3. Iyer, N., Jayanti S., Lou K., Kalyanaraman Y., Ramani K.: Three-Dimensional Shape Searching: State-Of-The-Art Review and Future Trends, Computer Aided Design, Vol. 37(5),(2005) 509-530
4. Chen, Y., Wang, J. Z., Krovetz, R.: CLUE: Cluster-based Retrieval of Images by Unsupervised Learning, IEEE Transactions on Image Processing, Vol. 14(8), (2005), 1187-1201
5. Sheikholeslami, G., Chang, W., Zhang, A.: SemQuery: Semantic Clustering and Querying on Heterogeneous Features for Visual Data, IEEE Transactions on Knowledge and Data Engineering (TKDE), Vol. 14(5), (2002) 988-1002.
6. Jing, F., Li, M., Zhang, H-J., Zhang, B.: A Unified Framework for Image Retrieval Using Keyword and Visual Features, IEEE Transactions on Image Processing, Vol. 14(7), (2005) 979 – 989
7. Swift K.G., Booker, J.D.: Process Selection: From Design to Manufacture, (1998), John Wiley and Sons, NY
8. Ip, Y., Regli, W. C.: Manufacturing Processes Recognition of Machined Mechanical Parts Using SVMs. AAAI2005, (2005) 1608-1609
9. Ip, Y., Regli, W. C.: Content-Based Classification of CAD Models with Supervised Learning, Computer Aided Design and Application, Vol.2 (5), (2005) 609-618
10. Zhang, C., Chen, T.: A New Active Leaning Approach for Content-Based Information Retrieval, IEEE Transactions on Multimedia Special Issue on Multimedia Database, Vol. 4(2),(2002) 260-268
11. Barutcuoglu, Z., Decoro, C.: Hierarchical Shape Classification Using Bayesian Aggregation, Shape Modeling International, (2006)

12. Giacinto, G., Roli, F: Ensembles of Neural Networks for Soft Classification of Remote Sensing Images, European Symposium on Intelligent Techniques, (1997) 166-170
13. Yan, R., Hauptmann, A., Jin, R., Liu, Y.: On Predicting Rare Class with SVM Ensemble in Scene Classification, IEEE International Conference on Acoustics, Speech and Signal Processing, (2003)
14. Senior, A.: A Combination Fingerprint Classifier, IEEE Transactions on Pattern Analysis and Machine Intelligence, Vol. 23(10), (2001) 1165-1174
15. Xu, L., Krzyzak, A., Suen, C.Y.: Methods of Combining Multiple Classifiers and Their Applications to Handwriting Recognition. IEEE Transactions on Systems, Man and Cybernetics, Vol. 22(3), (1992) 418-435
16. Roli, F., Kittler, J., Windeatt, T. (eds.): Multiple Classifier Systems, Lecture Notes in Computer Science, Vol. 3077, (2004)
17. Fumera, G., Roli, F.: A Theoretical and Experimental Analysis of Linear Combiners for Multiple Classifier Systems. IEEE Transactions on Pattern Analysis and Machine Intelligence, Vol. 27(6), (2005) 942-956
18. Kittler, J., Hatef, M., Duin, R., Matas, J.: On Combining Classifiers, IEEE Transactions on Pattern Analysis and Machine Intelligence, Vol. 20(3), (1998) 226–239
19. Tax, D., Breukelen, M.V., Duin, R., Kittler, J.: Combining Multiple Classifiers by Averaging or by Multiplying? Pattern Recognition, Vol. 33, (2000) 1475--1485
20. Kuncheva, L.I..: A Theoretical Study on Six Classifier Fusion Strategies," IEEE Transactions on Pattern Analysis and Machine Intelligence, Vol. 24, (2002) 281-286
21. Verikas, A., Lipnickas, A., Malmqvist, K., Bacauskiene, M., Gelzinis, A.: Soft Combination of Neural Classifiers: A Comparative Study, Pattern Recognition Letters, Vol. 20(4), (1999) 429-444
22. Benediktsson, J.A., Sveinsson, J.R., Ersoy, O.K., Swain, P.H: Parallel Consensual Neural Networks, IEEE Transactions on Neural Networks, Vol. 8(1), (1997) 54-64
23. Hashem, S.: Optimal Linear Combination of Neural Networks, Neural Networks, Vol. 10, (1997) 599-614
24. Ueda, N.: Optimal Linear Combination of Neural Networks for Improving Classification Performance, IEEE Transactions on Pattern Analysis and Machine Intelligence, Vol.22 (2000) 207-215
25. Pu, J., Ramani, K.: A 3D Model Retrieval Method Using 2D Freehand Sketches. Lecture Notes in Computer Science, Vol. 3515 (2005) 343-347
26. Jayanti, S., Kalyanaraman, Y., Iyer, N., and Ramani, K.: Developing an Engineering Shape Benchmark for CAD models, Special Issue on Shape Similarity Detection and Search for CAD/CAE Applications, Journal of Computer Aided Design, Vol.38(9), (2006), 939-953
27. Hou, S., Ramani, K.: Sketch-based 3D Engineering Part Class Browsing and Retrieval, EuroGraphics Symposium Proceedings on Sketch-Based Interfaces and Modeling, 131-138, (2006),
28. Kuncheva, L.I.: Combining Pattern Classifiers: Methods and Algorithms Hoboken, (2004), N.J.: Wiley,
29. Pu, J., Jayanti S., Hou, S., Ramani, K.: Similar 3D model retrieval based on multiple level of detail, accepted by the 14th Pacific Conference on Computer Graphics and Applications, (2006)
30. Hsu, C. W., Change C.C., Lin, C. J.: A Practical Guide to Support Vector Classification, (2004) http://www.csie.ntu.edu.tw/~cjlin/libsvm
31. Platt., J.: Probabilistic Outputs for Support Vector Machines and Comparison to Regularized Likelihood Methods, in Advances in Large Margin Classifiers, Smola, A., Bartlett, P., Schoelkopf, B., Schuurmans, D. (eds.), (1999) 61-74

A Bayesian Network Approach to Multi-feature Based Image Retrieval*

Qianni Zhang and Ebroul Izquierdo

Department of Electronic Engineering, Queen Mary, University of London
London, U.K.
{qianni.zhang, ebroul.izquierdo}@elec.qmul.ac.uk

Abstract. This paper aims at devising a Bayesian Network approach to object centered image retrieval employing non-monotonic inference rules and combining multiple low-level visual primitives as cue for retrieval. The idea is to model a global knowledge network by treating an entire image as a scenario. The overall process is divided into two stages: the initial retrieval stage which is concentrated on finding an optimal multi-feature space stage and doing a simple initial retrieval within this space; and the Bayesian inference stage which uses the initial retrieval information and seeks for a more precise second- retrieval.

1 Introduction

The general problem of retrieving, classifying and recognizing patterns in images has been investigated for several decades by the image processing and computer vision research communities. Learning approaches, such as neural networks, kernel machines, statistical and probabilistic classifiers, can be trained to obtain satisfactory results for very specific applications. If the structure of the database and relevant low-level features are known then the constrained pattern recognition problem can be solved with relatively high accuracy. Much of the related work on image classification for indexing and retrieval has focused on the definition of low-level descriptors and the generation of metrics in the descriptor space [1], [2]. These techniques are aimed at defining image signatures using primitives extracted from the content, e.g. pixel patterns and dynamics in images and video or sampling patterns in audio signals. These descriptors are extremely useful in some generic image classification tasks or when classification based on query by example is considered. However, if the aim is to annotate single objects in complex images using semantic words or sentences, there are two questions to be answered in order to solve difficulties that are hampering the progress of research in this direction. Firstly, how to deal with the subjective interpretation of images by different users under different conditions? Secondly, how to link semantically meaningful objects in images with low-level metadata?

The first difficulty, originated from the dependence of perceptual similarity on both user and context, can be tackled by applying rule based semantic reasoning, typically expressed as ontology, to facilitate the recognition of concepts based on experts' rules and experience. The common solution to the second difficulty considers linking a

---

* The work leading to this paper was partially supported by the European Commission under contracts FP6-001765 aceMedia and FP6-027026 K-Space.

Y. Avrithis et al. (Eds.): SAMT 2006, LNCS 4306, pp. 138–147, 2006.
© Springer-Verlag Berlin Heidelberg 2006

semantic concept, e.g., a keyword, with low-level metadata. In order to achieve this, a machine needs to learn associations between complex combinations of low-level patterns and conceptual objects. Consequently, complex combinations of features building high-dimensional and heterogeneous feature spaces need to be considered. Therefore aim of this paper is two-fold: to devise a Bayesian Network approach to object-centered image retrieval employing non-monotonic inference rules and to combine multiple visual primitives as cue for retrieval. In the literature the Bayesian approach is mainly used in region or object-based image retrieval systems [3], [4], [5], [6], in which the object's likelihood can be calculated from the conditional probability of feature vectors. These systems use probabilistic reasoning to exploit the relationship between objects and their features. Some other systems employ Bayesian approach in scenario of scene classification e.g., sunset, indoor, outdoor, landscape [7], [8]. However, this kind of classifications has been restricted to mutually exclusive categories, and so is only suitable to images that have only on dominant concept. But in more realistic scenarios in image classification and retrieval, images are complex and usually consist of many semantically meaningful objects. Therefore the relationships between semantically meaningful concepts of the objects cannot be ignored and need to be explored in great degree. In [9], multi-categories are introduced with a simple "sub-class-of" relationship between some of the parent categories with their children categories. While in [10], [11], a statistical analysis of the relationships among concepts is adopted and achieve image retrieving by using metadata (e.g. the RDF triples or semantic web ontology model) that describe and organize the concepts in images.

In this paper, the idea is to model a global knowledge network, thus one entire image is treated as a scenario, and the beliefs is formulated regarding concepts in possibly small regions of the entire image. However in our current work, segmentation is not assumed since segmenting an image into single object is almost as challenging as the semantic gap problem itself. A simple approach is taken to deal with objects in images small image blocks of regular size are considered, which are referred to as "elementary building blocks" in the following of the paper. The overall process is divided into two stages: the initial retrieval stage and Bayesian belief network inference stage, as shown in Fig.1. The initial retrieval stage is concentrated on finding an optimal multi-feature space based on a set of well-selected "representative blocks" for each concept, and then a simple initial retrieval is done within this space, providing essential information for the second stage. In this step the Multi-Objective Optimization (MOO) technique is adopted for estimating the 'optimal' multi-feature metric space [12], [13], [14]. The Bayesian inference stage models initial believes on probability distributions of concepts from the initial retrieval information and construct a Bayesian belief network. A more precise second- retrieval is conducted by inferring the presence of objects from the interactions between different concepts on image level.

This paper is organized as follows: Section 2 describes how the multi-feature space is constructed using MOO technique, and how the initial block-level retrieval is done. Section 3 introduces the method of building the Bayesian belief network in this scenario and inferring posterior probabilities out of the network. Experimental results are shown in Section 4 and the paper concludes with Section 5.

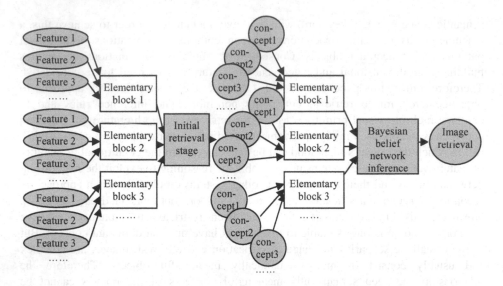

Fig. 1. Overview of the two-stage retrieval process

2 Initial Retrieval in an Optimized Multi-feature Space

Retrieval of an image relies on the retrieval of the objects within the image. The considered objects in an image are in the following analysis represented by blocks [14]. An example of one image being divided into elementary building blocks that contain different concepts is illustrated in Fig. 2.

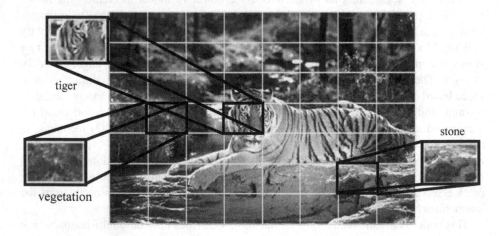

Fig. 2. Example of an image being divided into some elementary building blocks for concepts such as 'tiger', 'vegetation' and 'stone'

2.1 Feature Extraction and Distance Calculation

The primitives used by the proposed analysis are selected from the visual descriptors including MPEG-7 Colour Layout (CLD), Colour Structure (CSD), Dominant Colour (DCD), and Edge Histogram (EHD) [1]. Two texture features are also used as low-level primitives: Texture feature based on Gabor Filters (GF) [15] and Grey Level Co-occurrence Matrix (GLCM) [16]. Additionally, to emphasis invariance to saturation, Hue-Saturation- Value (HSV) [17] color system is also considered.

Let $S = \{s^{(i)} \mid i = 1,...,m\}$ be the training set of elementary building elements. For n low-level descriptors, a $m \times n$ matrix is formed in which each element is a descriptor vector. The centroid for each descriptor is calculated by finding the block with the minimal sum of distances to all other blocks in S. All the centroids across different descriptors form a particular set of vectors $\bar{s} = \{\bar{v}_1, \bar{v}_2, ,..., \bar{v}_n\}$, in which $\bar{v}_i$ is the centroid vector for all the vectors of the ith descriptor used. In general $\bar{s}$ does not necessarily represent a specific block of S. Taking $\bar{s}$ as an anchor, distances from all other elementary blocks to the centroid in each feature space can be estimated using:

$$d_j^{(i)} = dist(\bar{v}_j, v_j^{(i)}) \tag{1}$$

Thus, for a given concept representing an object the matrix

$$
\begin{matrix}
d_1^{(1)} & d_2^{(1)} & \cdots & d_m^{(1)} \\
d_1^{(2)} & d_2^{(2)} & & d_m^{(2)} \\
\vdots & & \ddots & \vdots \\
d_1^{(n)} & d_2^{(n)} & \cdots & d_m^{(n)}
\end{matrix}
\tag{2}
$$

is built. In (2) each row contains distances of different descriptors estimated for the same block, while each column display distances for the same descriptor for all blocks.

2.2 Combining Features Using Multi-objective Optimization

To combine the distance calculated from matrix (2), the most straightforward candidate of possible metrics in the multi-feature space is the linear combination of the distances defined for the descriptors:

$$m(A, D) = \alpha_1 d_1 + \alpha_2 d_2 + \alpha_3 d_3 + ... \tag{3}$$

Here A is the set of weighting factors and D is the set of distance functions for single descriptors. The problem of finding the suitable metric consists of finding the optimal set of weighting factors α, where optimality is regarded in the sense of both concept representation and discrimination power. Please note the optimality here means that, under the assumed model, there does not exist any other combination metric which leads to a lower retrieval error.

This optimization problem can be tackled by minimizing one or several objective functions as in (4). Due to the complexity of the visual descriptions of natural semantic concepts, it is not possible to find an image or image block that can represent a concept completely. The representation does not have to only contain all the visual features that are representing the objects of such concept, but also these visual features have to discriminate the concept from other semantic elements. Considering this, a group of 'representative building blocks', usually consisting 10 to 20 elements are selected to represent a concept. Each block in the representative group is used as an objective function.

Assume a database is built up in which all images are split into blocks and the visual features of the blocks are extracted. Given a semantic concept that the user would like to retrieve, the first step of the proposed approach is to build up a training group of 'representative building blocks'. From this training group the machine will learn the underlying visual descriptive nature of the concept. The number of the blocks in such a selection from the database can vary according to the subjective observation of the professional users. These blocks should be showing what features the user think is most representative for the concept.

Using the visual features of the training blocks, the centroid can be calculated as described in 2.2, and the distance matrix (2) can be built. For a given semantic concept and its according distance matrix (2), the optimization of (3) is then performed on the objective functions set as:

$$M(A,D) = \begin{cases} m_1(A, D^{(1)}) \\ m_2(A, D^{(2)}) \\ .... \\ m_m(A, D^{(m)}) \end{cases} \tag{4}$$

In (4) A is the collection of weighting factors, and $D^{(i)}$ is the distance vector of the ith block. The optimal solution is to find the minimal value of M and its corresponding $A = \{\alpha_j \mid j = 1,...,n\}$, subject to constraint $\sum_{j \in (1,n)} \alpha_j = 1$.

This set of weighting factors α is assumed to be the metric that represents the symbolic nature of the concept within a multi visual feature space. The initial retrieval is done in this space and if any elementary block of an image is classified as relevant, the entire image is classified as relevant. The results in this stage are used for constructing the belief ontology in Bayesian networks.

3 Bayesian Network Inference

Bayesian methods are a fundamental approach to image analysis, computer vision and pattern recognition problems [18], [19]. In this work the decisions are inferred using Bayesian networks that are conventional directed acyclic graphs with conditional probability distributions linking the "causes" to the "effects". All the probabilities used in the Bayesian Network are computed from information in the belief ontology which is created using the initial retrieval results.

For a particular concept user has in mind, each image in database can be classified into two classes: "relevant" or "irrelevant". The two possible classes are denoted as C_k, where $k \in 1,2$ (in this paper C_1 corresponds to relevant and C_2 corresponds to irrelevant). If probabilities are denoted as $P(\cdot)$, the prior probability of class membership is denoted as $P(C_k)$. The features used to help the inference are denoted as a set F, and $P(\mathbf{F})$ is the evidence factor. In this Bayesian framework, all inferences are based on the posterior probability function $P(C_1 | \mathbf{F})$, which is obtained by combining the class-conditional observation models with the class prior probability according to Bayes law:

$$P(C_1 | \mathbf{F}) = \frac{P(\mathbf{F} | C_1)P(C_1)}{P(\mathbf{F})} \qquad (5)$$

The classification criterion used is the most common *maximum a posteriori* (MAP) in Bayesian classification problems. It is given by:

$$C_1 = \arg\max P(C_1 | \mathbf{F}) \qquad (6)$$

In a two-class Bayesian classification, each image is classified as relevant if its posterior probability is greater than some threshold.

3.1 The Belief Ontology and Bayesian Network

In the belief ontology network, each node represents a concept such as "bear" or "honey" from the vocabulary of the domain of application. In this paper the belief ontology is modeled using Bayesian belief network because they have similarity that the nodes represent propositions which are either true or false and has probabilities associated with co occurrence relationships. However, the co occurrence relationships between concepts are not causal and the probabilities kept with these relationships simply measure statistical association.

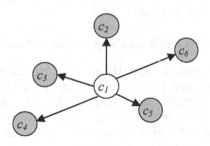

Fig. 3. Example of a part of Bayesian network centered on concept c_1

On one hand, because a Bayesian network is naturally capable of encoding the joint probability distribution, it is considered as a representation of ontology. One the other hand, because it produces the posterior join probability distribution based on various evidences, it is also an inference engine that can exploit information contained in interrelationships and dependencies between elements.

3.2 Constructing the Bayesian Networks

In order to explore relationships between these semantically meaningful concepts of the objects in images, a small ontology containing concepts of objects that are typical in the experimental database is first pre-defined. Each concept in the belief ontology network is considered one by one for each elementary block in images in database. As described in Section 2, the initial labeling of elementary blocks is done within an optimized multi-feature space. The results obtained directly from this step are shown in Section 4.

For each concept in the belief ontology containing p concepts $c_t, t = 1,..., p$, a Bayesian network is constructed by considering this concept c_1 and all other concepts that are directly linked to c_1. For instance, when the scenario is to search for images containing one particular concept c_1 that user has in mind, C_k $k \in 1,2$ represents that either an image is relevant to c_1 ($k = 1$), or is irrelevant to c_1 ($k = 2$). A Bayesian network such as the one shown in Fig. 3 can be constructed.

However not all concepts in the ontology need to be included in a Bayesian network, since each Bayesian network centered on a concept only focuses on a part of the belief ontology including only the immediately related concepts associated with the center concept.

4 Experimental Setup and Evaluation

The test data contains 700 images selected from 'Corel' dataset. The images are labeled manually on 5 predefined concepts as ground truth. The concepts are "building" (141), "cloud" (264), "grass" (279), "lion" (100), and "tiger" (100), where the numbers in brackets after concepts are the numbers of images of the concepts in database according to the ground truth.

4.1 Experiments of Direct Retrieval Using Obtained Multi-feature Metric

Initially a group of 10 positive and 10 negative representative blocks for each concept is manually selected per concept. Using both kinds of training blocks, optimized multi-feature metrics are obtained. In order to show the improved retrieval performance of using multiple features over retrieval using single descriptors, an accuracy value of doing completely the same retrieval process but using only each of the 7 used single descriptors are also computed and listed in the table for comparison.

Table 1. Accuracy values of retrieval directly using obtained metrics compared with using single descriptors

%	Obtained metric	CLD	CSC	DCD	EHD	GF	GLCM	HSV
building	70	48	24	20	**74**	40	38	42
cloud	**79**	76	70	38	68	28	34	78
grass	**92**	92	86	28	82	64	88	88
lion	**88**	50	36	16	50	24	40	66
tiger	**60**	2	46	7	14	26	34	57

As shown in experiments results in Table 1, the proposed approaches using optimized multi-feature metrics generally perform better than the retrieval based on single descriptors. Even though in some cases specific single descriptors are dominant for a concept, the result from proposed approach is very close to it.

4.2 Building Bayesian Inference Network Based on Initial Retrieval Results

The initial retrieval results from experiments presented in Section A are used as sources for estimating variables and building the Bayesian belief network model.

Table 2. Accuracy values of retrieval using Bayesian network compared with initial results in initial experiment

%	building	cloud	grass	lion	tiger
Initial results	70	79	92	88	**60**
Bayesian net	**72**	**84**	**94**	**92**	**60**

This set of experiments show that retrieval results using posterior probabilities inferred from Bayesian networks are always more accurate comparing with the direct retrieval results in multi-feature space. The only exception is for concept 'tiger', the accuracy remains the same as 60%. Interestingly, it can be observed from our experiments that, in general, if the initial retrieval result is higher than 60%, the Bayesian inference brings good enhancements to the result. But for low-accuracy initial retrieval results, such as lower than 50%, the Bayesian approach does little improvements. This is because the inaccuracy in estimating the prior probability or the likelihood introduces much noise into the belief network constructed.

5 Conclusions and Future Work

A Bayesian network-based framework for object-centered image retrieval is presented. In our method, first the images are decomposed into blocks to enable a real object-centered analysis. By using Multi-Objective Optimization on a group of key-representative image blocks, an optimal similarity metric per semantic concept is obtained. Bayesian inference is then performed to map the retrieval from block level back to image level. The results, though largely depending on the choice of representative blocks in the first stage, show definite improvements by the Bayesian inference step.

In future work experiments involving more concepts on bigger dataset will be done. By doing so we hope the potential ability of Bayesian network applied in our scenario can be further explored.

References

[1] S.-E Chang, T Sikora, A. Purl, "Overview of the MPEG-7 Standard", IEEE Transactions on Circuits and Systems for Video Technology, vol. 11, No. 6, pp. 688-695, 2001.
[2] Mojsilovic, "A computational model for color naming and describing color composition of images", IEEE Transactions on Image Progressing, vol 14, No. 5, pp. 690-699, 2005.
[3] L. Fei-Fei, R. Fergus, and P. Perona, "A Bayesian Approach to Unsupervised One-Shot Learning of Object Categories", Proceedings of the Ninth IEEE International Conference on Computer Vision (ICCV'03)
[4] D. Hoiem, R. Sukthankar, H. Schneiderman, L. Huston , "Object-Based Image Retrieval Using the Statistical Structure of Images", IEEE Conference on Computer Vision and Pattern Recognition, June, 2004.
[5] R. Fergus, P. Perona, A. Zisserman, "Object Class Recognition by Unsupervised Scale-Invariant Learning", IEEE CVPR '03, 2003
[6] Scott Helmer, David G. Lowe, "Object Class Recognition with Many Local Features", IEEE CVPRW'04, 2004
[7] Aditya Vailaya, Mario Figueiredo, Anil Jain, Hong Jiang Zhang, "A Bayesian Framework for Semantic Classification of Outdoor Vacation Images" in Proc. SPIE: Storage and Retrieval for Image and Video Databases VII, vol. 3656, pp. 415-426, San Jose, CA, January, 1999
[8] J. Luo, J., Savakis, A., "Indoor vs Outdoor Classification of Consumer Photographs Using Low-level and Semantic Features", IEEE, ICIP01(II: 745-748).
[9] Jia Li, James Z. Wang, "Automatic Linguistic Indexing of Pictures by a Statistical Modeling Approach", IEEE Transactions On Pattern Analysis and Machine Intelligence, Vol. 25, No. 9, 2003
[10] Xiaofei He, Oliver King, Wei-Ying Ma, Mingjing Li, and Hong-Jiang Zhang, "Learning a Semantic Space From User's Relevance Feedback for Image Retrieval", IEEE Transactions On Circuits And Systems For Video Technology, Vol. 13, No. 1, January 2003
[11] W. H. Adamsy, G. Iyengary, C-Y Linz, M. R. Naphadez, C. Netiy, H. J. Nocky, J. R. Smithz, "Semantic Indexing of Multimedia Content using Visual, Audio and Text cues", EURASIP JASP 2003
[12] R.E, Steuer, Multiple criteria optimization. Theory, Computation, and Application. New York: Wiley 1986.
[13] J. Knowles, D. Corne, "Approximating the non-dominated front using the Pareto Archived Evolution Strategy", 1999.
[14] Q. Zhang and E. Izquierdo "A Multi-Feature Optimization Approach to Object-Based Image Classification", CIVR 2006.

[15] B.S. Manjunath, W.T. Ma, "Texture features for browsing and retrieval of image data," IEEE Trans. On Pattern Analysis and Machine Intelligence, vol. 18, no. 8, pp. 837-842, August 1996.

[16] M. Tuceryan and A. K. Jain, Texture Analysis. The Handbook of Pattern Recognition and Computer Vision (2nd Edition), pp. 207-248, World Scientific Publishing Co., 1998.

[17] M. Swain, and D. Ballard, "Color indexing," International Journal of Computer Vision, 1991.

[18] R. O. Duda, P. E. Hart and D. G. Stork, Pattern Classification. John Viley & Sons, CA, 2001

[19] J. Pearl, Probabilistic Reasoning in Intelligent Systems: Networks of Plausible Inference. Morgan Kaufmann Publishers, Inc, 1998.

Extraction of Motion Activity from Scalable-Coded Video Sequences*

Luis Herranz, Fabricio Tiburzi, and Jesús Bescós

Grupo de Tratamiento de Imágenes, Escuela Politécnica Superior
Universidad Autónoma de Madrid, E-28049 Madrid, Spain
{Luis.Herranz, Fabricio.Tiburzi, J.Bescos}@uam.es

Abstract. This work presents an efficient approach for the calculation of the MPEG-7 descriptor for motion activity from scalable-coded video sequences, which include scalable motion vectors and variable block sizes. We first describe the adaptation of the constant block-size assumption of the MPEG-7 descriptor to this new coding domain. Then we compare the results obtained with those for MPEG-1 videos in the context of a common application of this descriptor: video summarization. The comparable quality of these results and the gain in efficiency support the presented approach.

Keywords: motion activity, scalable video, video indexing, video analysis.

1 Introduction

Motion activity is a perceptual feature which aims at describing the amount of action that shows a video segment. This is useful for tasks such as video indexing or analysis. Therefore, the problem of motion activity extraction has been widely addressed in the last years and some of the existing approaches have been included into standards. This is the case of the MPEG-7 descriptor for motion activity.

New coding paradigms such as H.264/AVC[1] and scalable video coding[2] intend to address the requirements of the evolving multimedia scenarios. Works that extend the existing approaches to obtain motion activity from sequences coded in this new ways are relatively scarce. In this paper we focus on extracting this feature from scalable-coded videos and to comparatively evaluate the results obtained respect to those achieved on MPEG-1/2 videos.

A common application of motion activity is to guide key frame selection for the generation of video summaries. For this reason, we have used a distortion measure on video summaries to compare motion activity extraction for MPEG video and scalable video.

The rest of the paper is organized as it follows. Sections 2 and 3 provide brief descriptions of the concept of motion activity and of the scalable video coding

---

* Work partially supported by the European Commission under the 6th Framework Program (FP6-001765 - aceMedia Project) and by the Spanish Government under Project TIN2004-07860-C02-01.

Y. Avrithis et al. (Eds.): SAMT 2006, LNCS 4306, pp. 148–158, 2006.
© Springer-Verlag Berlin Heidelberg 2006

framework studied. The method to extract motion activity from scalable-coded sequences is presented in Section 4. Section 5 presents the approach used to select key-frames from motion activity. In Section 6 we evaluate and compare the summaries obtained in terms of distortion respect to original sequence. Finally, experimental results and conclusions are given in Sections 7 and 8.

2 Motion Activity

Motion activity is defined as a measure of the "intensity of motion" or "pace" of a sequence such as it is perceived by a human. Although it depends on events such as object or camera motion, as it occurs with many other subjective or perceptual elements there are no objective criteria that completely determine it. In order to somehow estimate motion activity MPEG-7[3] provides a descriptor[4] defined as the standard deviation of the motion compensation vectors, which are directly extracted from the P frames of MPEG-1/2 coded bitstreams. This deviation is then normalized by the frame resolution and by the P frame rate, and it is finally quantized into 5 levels.

3 A Wavelet-Based Scalable Video Codec

Scalable video coding aims at coding a video sequence so that a single encoded bitstream can be efficiently decoded at different fidelity levels. This reduces content adaptation to almost a selection of the necessary parts of the bitstream. There are many alternatives to provide scalability in video coding, such as layered approaches and embedded coding approaches. In this work we focus on a fully scalable video codec using Motion Compensated Temporal Filtering (MCTF). However, the technique is generic and easy to extend to other approaches such as hierarchical B frames in the recent standard MPEG-4 SVC[5].

Wavelet coding enables a natural multiresolution framework for highly scalable video coding[2]. Most works on this subject are based on two steps: first, a 2D spatial discrete wavelet transform (2D DWT); second, another wavelet transform in the temporal axis, combined with motion compensation (MCTF). There are many codecs that use the t+2D framework, where the temporal transform is performed before the spatial transform[6][7].

In this paper we consider the scalable coding approach based on the t+2D framework described in [8]. It uses MCTF with hierarchical variable size block matching (HVSBM)[9] and forward motion compensation. It considers spatial, temporal and quality (SNR) scalability. In this paper, we will refer to this specific codec as SVC (for scalable video codec).

When dealing with video sequences coded in this way, the adaptation process is a simple discarding process of the not required information, such as spatial or temporal subbands, or wavelet coefficient bits for bitrate scalability. In the decoder, the encoding process is reversed, as shown in Fig. 1. First, wavelet coefficients of the required subbands are entropy decoded, and an inverse 2D DWT is performed to obtain the pixels of the frames. Then, inverse MCTF reconstructs the frames in the GoP at the required temporal resolution. Note that only entropy decoding of the motion vectors is necessary to extract these motion vectors.

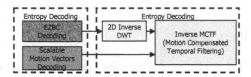

Fig. 1. Scheme of the decoder

3.1 Hierarchical Variable Size Block Matching

Most established video standards, such as MPEG-1 and MPEG-2, use a constant block size block matching algorithm to perform motion estimation for compensation. Thus, only the motion vectors corresponding to each block need to be transmitted.

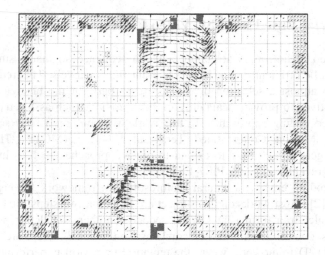

Fig. 2. Example of the motion field that results from applying HVSBM with the scalable codec

Variable size block matching can represent more effectively the motion information using less motion vectors in big areas with uniform motion. In HVSBM, the motion field is partitioned in a quadtree structure, and a motion vector is assigned to each block in each leaf block. The partition structure is transmitted together with the motion vectors. This scheme for motion estimation is used in many video codecs, such as the recent AVC/H.264[10] and most of the MCTF frameworks for scalable video coding. Fig. 2 shows an example of the quadtree structure and the motion field obtained using HVSBM.

In scalable video coding using MCTF, each temporal decomposition level includes a set of motion vectors coded together with the residual coefficients. This information is enough to invert the transform.

4 Motion Activity Extraction

Temporal scalability on SVC is obtained performing successive dyadic decompositions in the temporal axis, using MCTF, which is based on motion vectors. These vectors can be used to compute motion activity in a similar way to that followed for MPEG-1/2. However, there are two main differences: the temporal scalability itself and the variable block size for motion compensation.

4.1 Motion Activity Levels

Motion vectors in SVC are organized to enable temporal scalability: depending on the temporal level, the temporal distance (measured in frames of the original sequence) between the actual frame and the reference frame is different.

Each set of motion vectors corresponding to each temporal level represents the same motion information of the GoP, but at different scales. This is important in terms of efficiency, as the number of motion vectors to process per GoP depends on the desired level, being less in lower levels at the expenses of reducing precision in the temporal axis.

For instance, consider a GoP of 8 frames with three temporal decompositions (see Fig. 3). Three sets of motion vectors are necessary to decode the sequence at full frame rate (level $k=3$). Motion vectors from the first, second and third level refer to temporal distances of 4, 2 and 1 frames. Thus, if we directly computed the motion activity we would obtain 1, 2 and 4 values per GoP, respectively.

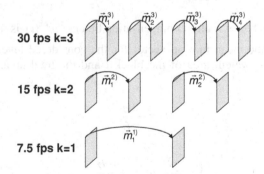

Fig. 3. Levels of motion vectors in SVC for a GoP of 8 frames. $m_i^{k)}$ corresponds to the ith vector of the kth level.

4.2 Dealing with Variable Block Size

As opposed to the constant (macro)block size used for motion compensation in MPEG-1/2, block sizes in SVC are variable. This further prevents from directly using

SVC motion vectors to calculate the MPEG-7 descriptor, which assumes that all the blocks have the same size, that is, that the motion field is uniformly distributed. A solution is to include the size of the block in the computation of the motion activity. In fact, it's possible to construct an equivalent motion vector grid where all the vectors are referred to the motion of constant size blocks. Fig. 4 shows a diagram of this conceptual representation where the grid consists of 16x16 pixels macroblocks. In this equivalent grid, motion activity can be computed as described in the MPEG-7 descriptor for MPEG-1/2.

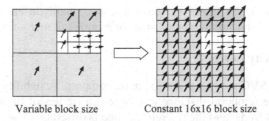

Variable block size   Constant 16x16 block size

Fig. 4. Equivalent motion vectors in 16x16 blocks

Note that this last replication is only conceptual and actually there is no real need to obtain it: we can just include the relative block size into the expression of the descriptor as a weighting factor, hence even much reducing the number of operations required (respect to the constant block size situation). In conclusion, the motion activity at temporal level k can be calculated as:

$$I^{k)} = \sqrt{\frac{1}{A_{MB}} \sum_{i=0}^{N-1} a(i) \left\| \vec{m}^{k)}(i) - \bar{\vec{m}}^{k)} \right\|^2} \tag{1}$$

where N is the number of motion compensation blocks, $\vec{m}^{k)}(i)$ is the motion vector at block i, and $\bar{\vec{m}}^{k)}$ the mean motion vector for the considered level; $a(i)$ and A_{MB} are respectively the equivalent area of the block i and the total area, both measured in macroblocks:

$$a(i) = \frac{A(i)}{256} \tag{2}$$

$$A_{MB} = \sum_{i=0}^{N-1} a(i) = \frac{W \cdot H}{256} \tag{3}$$

$$\bar{\vec{m}}^{k)} = \frac{1}{A_{MB}} \sum_{i=0}^{N-1} a(i)\, \vec{m}^{k)}(i) \tag{4}$$

where $A(i)$ is the area of the block in pixels, and W and H are the frame dimensions.

The aforementioned efficiency gain depends almost solely on the number of involved motion vectors. If we compare with a 16x16 constant block size approach, this gain is the quotient between A_{MB} and the actual number of vectors N. Note that block sizes smaller than 16x16 can be also included with no conflict, just considering that their contribution to the total activity is a fraction of a block of 16x16 (e.g., a vector associated to a 8x8 block will contribute with a weight of 1/4).

5 Key-Frame Selection

5.1 Linear Sampling in Cumulative Motion Activity

Divakaran[11] proposes an interesting approach for key-frame extraction in MPEG videos using the motion activity feature. It assumes that, as motion activity measures inter frame change, its cumulative value should be a good indicator of content change. Hence, in order to extract n key-frames it divides the sequence into n "equal" parts by taking equal increments of the cumulative motion activity and then selecting key-frames at the middle of each segment. This linear sampling of the cumulative motion activity is aimed at ensuring that key-frames represent theoretically the same amount of information of the sequence.

In the original approach, the motion activity of P frames from the last GoP is averaged and assigned to I-frames, so that only these last frames can be selected as key-frames. This allows a fast extraction of the key-frames, as no motion compensation is required, but prevents from selecting any other frame as key-frame which may result in a degraded quality of the summary in GoPs with intense activity. In order to have a better temporal resolution in the summary, our algorithm also considers P frames as candidate key-frames.

Regarding to extending this approach to SVC, the frames from different temporal scales are considered as candidate key-frames. Also, due to the dyadic temporal decomposition, the pathway to decode a given frame in a GoP is usually shorter in SVC, as just a few references in the hierarchy have to be decoded. In contrast, in MPEG-1/2 is necessary to decode sequentially all the previous references, which for common GoP sizes often requires more motion compensation steps.

5.2 Composition of Inter-level Motion Activity

One problem of the scheme presented in the previous subsection is that it assumes continuous inter-frame prediction, as in fact this prediction is used to estimate the motion, and motion is used as a measure of information that grows progressively along the sequence. Although this is true in MPEG videos (except for I frames, where motion is replicated from the last P frame), in SVC each temporal level only includes motion vectors from odd to even frames (see Fig. 3), while there is no motion information for even to odd frame changes.

In order to maximize the number of frames covered by a continuous flow of activity along a whole GoP (to reach, at least, the MPEG situation) we have combined the activity values from different temporal levels. Thus, we also maximize the amount of change information that the motion activity measures. For instance, if we assumed

a GoP of 8 frames with three temporal decomposition levels, we would use the
following inter-level joint motion activity in a GoP (see Fig. 5):

$$I = I_1^{1)} + I_2^{2)} + I_4^{3)} \tag{5}$$

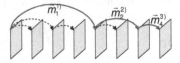

Fig. 5. Composition of level motion vectors in order to obtain an inter-level motion activity

Similarly, in the case of other temporal decompositions of the GoP the motion
activity usually can be added in a continuous flow using the value obtained from
different temporal levels, as all the frames in a GoP are related via motion vectors
from MCTF.

6   Video Summary Evaluation

As aforementioned, in order to test the performance of this approach for key-frame
selection, we have evaluated the quality of the resulting video summaries in terms of
similarity respect to the original sequence.

Though subjective evaluation would be required for proper evaluation, there are
several objective measures that can be used as indicators of the capability of a set of
key-frames to summarize a sequence. In [12] is introduced the semi-Hausdorff
distance between two sets of frames as a measure of similarity. This distance
measures how well all the frames of the original sequence are represented by the set
of the key-frames. In [13] is proposed a rate-distortion metric based on considering
the video summary as a quantized version of the original sequence, and then
measuring the distortion between the original sequence and the reconstructed one.
Note that the rate here refers to the rate of keyframes in the sequence and the
distortion is a global distortion metric of the whole set of keyframes respect to the
original sequence.

The semi-Hausdorff distance considers the summary as a set of frames instead of a
sequence of frames, hence ignoring the temporal relationship between frames. So, we
have considered it better to use the distortion approach.

Let $V = \{f_0, f_1, \ldots, f_{n-1}\}$ be a sequence with n frames, and let S be a subset or
summary of this sequence. A basic distortion measure of S respect to V could be
obtained by just applying:

$$D(S) = \max_{k \in [0, n-1]} d(f_k, f_k'), \quad f_k \in V, f_k' \in S \tag{6}$$

where missing frames in the summary are replaced by the nearest key-frame, and the
interframe distance $d(f_k, f_k')$ is the same proposed in [13], based on the Euclidean
distance in the Principal Components space.

7 Experimental Results

We have tested our algorithm with the *Stefan* (300 frames with a frame size of 352x240 pixels) and *Foreman* (300 frames with a frame size of 352x288 pixels) sequences. These are sequences that present segments with different intensities of motion activity. They were encoded both in MPEG-1 and SVC. For MPEG-1 we have used a typical GoP structure (IBBPBBPBBPBBPBB) at 30 frames per second. For SVC we have used 3 temporal decompositions, 3 spatial decompositions and 3 quality levels structured in GoPs of 8 frames. Consequently the SVC video included 3 sets of motion vectors, available with different temporal distances. The maximum block size for motion compensation was 64x64 pixels and the minimum 8x8 pixels.

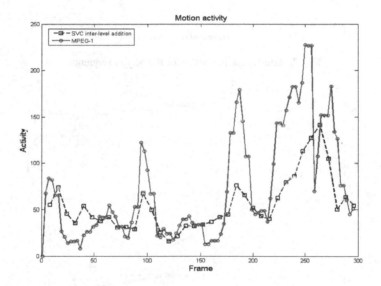

Fig. 6. Normalized motion activities for the Stefan sequence

Fig. 6 shows the motion activity for the Stefan sequence, obtained for MPEG-1 (in red) and for SVC (in dotted blue), here combining the three temporal levels as explained in section 0. Both curves follow approximately the activity present in the sequence, though the range of variation is wider in the case of MPEG-1.

For each sequence, a number of summaries were computed using the motion activity curves. The rate-distortion curves for each sequence are shown in Fig. 7 and Fig. 8, considering the number of key-frames as the rate. As expected, the distortion overall decreases with an increasing number of key-frames, but not monotonically (we are obtaining distortion from a curve of a feature, the motion activity, which is very subjective and related to the human perception).

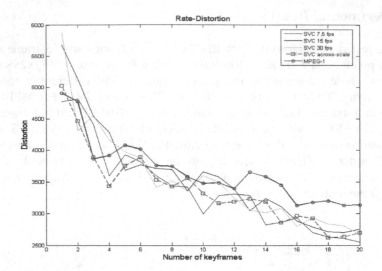

Fig. 7. Rate-Distortion curve for the *Stefan* sequence

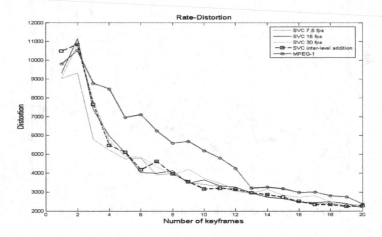

Fig. 8. Rate-Distortion curve for the *Foreman* sequence

A first observation is that all the distortion curves follow a similar shape, hence demonstrating the validity of our approach. A second one is that the MPEG-1 curve shows slightly more distortion than those obtained for SVC, and that the summary obtained using the joint inter-level motion activity is the most smooth, that is, the most correlated with the distortion measure. These observations are more evident in the *Foreman* sequence than in the *Stefan* one. Hence, we can conclude that the presented approach to obtain motion activity from scalable coded video is successful, and it achieves results better than those for MPEG-1, in terms of summary distortion using the proposed metric.

Comparing now the different distortion curves obtained for SVC, there are no significant differences among them in any of the two sequences. The inter-level addition to obtain a continuous flow of motion activity doesn't almost improve the results. Thus, the choice of a specific measure of motion activity for SVC should depend on other considerations, such as efficiency and availability of the motion vectors. In terms of efficiency, the use of the lowest temporal rate reduces the number of sets of motion vectors to one per GoP, which is enough for most of the applications, and allows a fast extraction of the motion activity descriptor.

8 Conclusions

In this paper we have presented an approach to extract the MPEG-7 descriptor for motion activity from scalable video sequences. The scalable coding framework used was based on temporal scalability of the motion vectors and on variable block-size motion compensation. These special domain characteristics have been taken into account to design an algorithm for obtaining a measure of the motion activity in a GoP.

The motion activity obtained for SVC has been compared with that obtained for MPEG sequences via a common application of this descriptor: key-frame selection for the generation of video summaries. Comparison has been carried out via a rate-distortion approach. The results show similar behavior in both domains, which validates the presented approach to efficiently compute this motion activity feature for the SVC domain.

Acknowledgments. This work resulted from the experience of a first author's stay at the Multimedia and Vision lab of the Queen Mary, University of London, under the supervision of Prof. Ebroul Izquierdo. We would like to acknowledge him and his two research assistants Marta Mrak and Nikola Sprljan for their support in the handling and understanding of motion compensation in their scalable video codec.

References

1. T. Wiegand, G.J. Sullivan, G. Bjntegaard, A. Luthra, "Overview of the H.264/AVC video coding standard", IEEE Transactions on CSVT, Vol. 13, No. 7, pp. 560- 576, Jul. 2003.
2. J.-R. Ohm, "Advances in Scalable Video Coding", Proc. of the IEEE, Vol. 93, Issue 1, Jan. 2005.
3. ISO/IEC 15938-3: 2002 Information technology. Multimedia content description interface – MPEG-7 Part 3: Visual.
4. S. Jeannin, A. Divakaran, "MPEG-7 visual motion descriptors", IEEE Transactions on CSVT, Vol. 13, No 6, pp. 720 - 724. Jun. 2001.
5. Heiko Schwarz, Detlev Marpe, and Thomas Wiegand, "Analysis of hierarchical B pictures and MCTF", IEEE International Conference on Multimedia and Expo (ICME'06), Toronto, Canada, Jul. 2006.
6. K. Shen and E. J. Delp, "Wavelet Based Rate Scalable Video Compression", IEEE Transactions on CSVT, Vol. 9, No. 1, pp. 109-122, Feb. 1999.

7. Peisong Chen, John W. Woods, "Bidirectional MC-EZBC with lifting implementation", IEEE Transactions on CSVT, Vol 14, No 10, pp. 1183-1194, 2004.
8. N. Sprljan, M. Mrak, G.C.K. Abhayaratne, E. Izquierdo, "A Scalable Coding Framework For Efficient Video Adaptation", Proc. 6th International Workshop on Image Analysis for Multimedia Interactive Services, WIAMIS 2005, Montreux, Switzerland, Apr. 2005.
9. M. Chan, Y. Yu, and A. Constantinides, "Variable size block matching motion compensation with applications to video coding", Proc. Inst. Elect. Eng., pt. I, Vol. 137, No. 4, pp. 205-212, Aug. 1990.
10. ITU-T Rec. H.264 & ISO/IEC 14496-10 AVC: Advanced Video Coding for Generic Audiovisual Services, 2003.
11. Divakaran, R. Radhakrishnan, K.A. Peker, "Motion activity-based extraction of key-frames from video shots", Proc. of ICIP'2002, Vol. 1, pp. 932-935, Sept. 2002.
12. H.S. Chang, S. Sull and S.U. Lee, "Efficient video indexing scheme for content-based retrieval", IEEE Transactions on CSVT, Vol. 9, No. 8, pp. 1269-1279, Dec. 1999.
13. Zhu Li, G.M. Schuster, A.K. Katsaggelos, "MINMAX optimal video summarization", IEEE Transactions on CSVT, Vol.15, No.10, pp. 1245- 1256, Oct. 2005.

Image Classification Using an Ant Colony Optimization Approach

Tomas Piatrik and Ebroul Izquierdo

Multimedia & Vision Research Group,
Queen Mary University of London
{tomas.piatrik, ebroul.izquierdo}@elec.qmul.ac.uk

Abstract. Automatic semantic clustering of image databases is a very challenging research problem. Clustering is the unsupervised classification of patterns (data items or feature vectors) into groups (clusters). Clustering algorithms usually employ a similarity measure in order to partition the database such that data points in the same partition are more similar than points in different partitions. In this paper an Ant Colony Optimization (ACO) and its learning mechanism is integrated with the K-means approach to solve image classification problems. Our simulation results show that the proposed method makes K-Means less dependent on the initial parameters such as randomly chosen initial cluster centers. Selected results from experiments of the proposed method using two different image databases are presented.

Keywords: Ant Colony Optimization (ACO), K-Means, Image Classification and Clustering.

1 Introduction

With the explosive growth of images in digital libraries, there is an increasing need for automatic tools to automatically annotate and organize image databases. Most works in automatic image annotation focused mainly on inferring high-level semantic information from low-level image features. Image recognition techniques facilitate the classification of images into semantically-meaningful categories and then label the images by the keywords that have been manually assigned to the categories. However, one common effort is to implement new techniques to improve inferring semantic information from low-level features in order to narrow the gap between low-level content based image description and their semantic counterparts: The semantic gap.

The fundamental step towards key-word based automatic image annotation is image classification. Automatic image classification is the task of classifying images into semantic categories with or without supervised training. Generally speaking there are two types of classification schemes: *supervised* and *unsupervised*. The task of supervised classification requires relevance feed-back and/or correction from a human annotator. On the contrary, unsupervised classification does not require human intervention. The main task of classification with unsupervised learning, commonly

Y. Avrithis et al. (Eds.): SAMT 2006, LNCS 4306, pp. 159–168, 2006.
© Springer-Verlag Berlin Heidelberg 2006

known as clustering, is to partition a given data set into groups (clusters) such that the data points in a cluster are more similar to each other than points in different clusters. Thus, the aim is to generate classes which allow us to discover similarities and differences, as well as to provide a concise summarization and visualization of the image content. A large number of classification algorithms have been developed in the past. Each of them has advantages and disadvantages [8]. However, efficient optimization techniques are able to reduce some undesired limitations. The optimization of these algorithms is currently the subject of active studies and promising results have started to emerge.

Recent developments in the optimization techniques are concentrated on natural and biological systems. These systems are inspired by the collaborative behavior of social animals such as birds, fish and ants and their formation of flocks, swarms and colonies. Some recent studies have pointed out that, the self-organization of neurons into brain-like structures, and the self-organization of ants into a swarm are similar in many respects [2]. Ants present a very good natural metaphor for evolutionary computation. With their small size and small number of neurons, they are not capable of dealing with complex tasks individually. The ant colony, on the other hand, can be seen as an "intelligent entity" for its great level of self-organization and the complexity of the tasks it performs. Their colony system inspired many researchers in the field of Computer Science to develop new solutions for optimization and artificial intelligence problems.

In this paper we present an approach to image partitional classification based on the Ant Colony Optimization (ACO) meta-heuristic. We start by briefly reviewing the K-Means algorithm in Section 2. Section 3 describes the characterizing aspects of the used ACO algorithm. In Section 4, the proposed Image Classifier algorithm is presented. Selected experimental results are given in Section 5. Conclusions and future work are discussed in Section 6.

2   The K-Means Algorithm

The K-Means algorithm is one of the most popular squared error based classification algorithm for partitioning N objects into K disjoint subsets S_j containing N_j objects [1]. Given a set of input patterns $X = \{x_1,...,x_i,...,x_N\}$, where $x_i = \{x_{i1},...,x_{if},...,x_{iD}\}^T \in \Re^D$ represents a feature vector and each measure x_{if} is said to be a feature. The aim of K-Means is the minimization of an objective function that is described by the following equation:

$$J = \sum_{i \in S_j}^{N_j} \sum_{j=1}^{K} d(x_i, c_j) , \qquad (1)$$

where $d(x_i, c_j)$ is a similarity (distance) measure between two objects (feature vectors). c_j is the cluster center of the objects in S_j. Thus, the criterion function J

attempts to minimize a chosen distance of each object from the center of the cluster to which the object belongs.

The algorithm consists of a simple procedure as follows:

1. Place K data points into the space represented by the objects that are being clustered. These objects represent an initial group of centroids.
2. Assign each object to the group that has the closest centroid
3. When all objects have been assigned, recalculate the positions of the K centroids.
4. Repeat Steps 2 and 3 until the centroids no longer move.

However, there are some limitations of this algorithm. K-means is sensitive to outliers and noise. Even if an object is quite far away from the cluster centroid, it is still forced into a cluster and, thus, distorts the cluster shapes. It also suffers from another drawback related to minimization of the objective function. In fact, since the algorithm uses randomly initial assignment of centroids, the "minimum" it reaches is not always a global minimum. Despite these limitations, the algorithm is used fairly frequently as a result of its simplicity and efficient implementations. For its improvement the K-means algorithm needs to be associated with some optimization procedures in order to be less dependent on given data and initialization.

3 Ant Colony Optimization

Ant algorithms were first proposed by Dorigo and colleagues [1, 3, 4] as a multi-agent approach to difficult combinatorial optimization problems like the traveling salesman problem (TSP) and the quadratic assignment problem (QAP). The characteristic of ACO algorithms is their explicit use of elements of previous solutions. The main underlying idea is loosely inspired by the behavior of real ants.

Compared to humans, an individual ant has very little brain power. The real power of ants resides in their colony brain. Ants are social insects, that is, insects that live in colonies and whose behavior is directed more to the survival of the colony as a whole than to that of a single individual component of the colony. The self-organization of those individuals is very similar to the organization found in brain-like structures, as indicated by Victorino Ramos [7]. An important and interesting behavior of ant colonies is, in particular, how ants can find the shortest paths between food sources and their nest.

One way ants communicate is by using chemical agents and receptors. For example, one ant is capable of distinguishing if another individual is a member of its own colony by the smell of its body. One of the most important of such chemical agents is the *pheromone*. Pheromones are molecules secreted by glands on the ant's body and once deposited on the ground, they start to evaporate. Ants use pheromone to communicate; one ant releases a molecule of pheromone that will influence the behavior of other ants. When one ant traces a pheromone trail to a food source, that trail will be used by many other ants that will reinforce that trail even more. Such autocatalytic processes will continue until a trail from the ant colony to the food source is established. Ants do not have the goal to create a trail that has shorter distance from nest to food source. Their goal is to bring food to the nest, but most of the time the pheromone trails they create are highly optimized.

In order to study in controlled conditions the ants foraging behavior, the binary bridge experiment has been set up by Deneubourg [2].

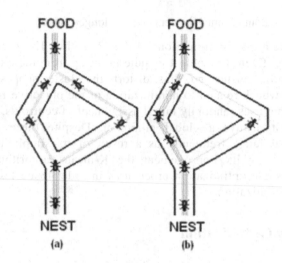

Fig. 1. Double bridge experiment

In Figure 1 are shown the experimental apparatus and the typical result of an experiment with a double bridge with branches of different lengths. As it can be seen in Figure 1a, at the beginning all ants choose their path on a 50 % probability. Ants that choose the shorter path get the food faster. Therefore, pheromone amount increases faster and makes shorter path more attractive to another ants (Figure 1b).

The Ant System algorithm (AS) was first inspired by the above learning mechanism and proposed to solving the Traveling Salesman Problem (TSP). Given a set of n cities and a set of distances between them, we call d_{ij} the length of the path between cities i and j. In the AS, the amount of pheromone in the trails is calculated using the following formula:

$$\tau_{ij}(t) = \rho \cdot \tau_{ij}(t-1) + \Delta \tau_{ij} .$$ (2)

The amount of pheromone τ in the edge(i,j) is calculated by first applying an evaporation coefficient ρ to the previous pheromone amount plus a new amount of pheromones calculated from all k ants that passed through the edge(i,j):

$$\Delta\tau_{ij} = \sum_{k=1}^{m}\Delta\tau_{ij}^{\;k} \;,$$

(3)

$$\Delta\tau_{ij}^{\;k} = \begin{cases} \dfrac{Q}{L_k}, \text{if } k \text{ - th ant use edge(i, j)} \\ 0 \qquad\qquad\text{, otherwise} \end{cases}.$$

(4)

Q is a parameter that specifies the amount of pheromones distributed by an ant k, and L_k is the tour length of ant k. It becomes clear that ants with minimum tour length will deposit a greater amount of pheromone to its path. Individual ants choose a next path based on this following formula:

$$p_{ij}^{\;k}(t) = \begin{cases} \dfrac{(\tau_{ij}(t))^{\alpha}(\eta_{ij})^{\beta}}{\sum\limits_{k\in allowed}(\tau_{ij}(t))^{\alpha}(\eta_{ij})^{\beta}} \;, \\ 0 \quad\text{, otherwise} \end{cases}$$

(5)

where $allowed_k(t)$ is the set of cities not visited by ant k at time t, and η_{ij} denotes a local heuristic which equal to $1/d_{ij}$ (and it is called 'visibility'). The parameter α and β control the relative importance of pheromone trail versus visibility.

The ACO algorithms became a promising research area and further improved algorithms as Ant-Q, Ant Colony System (ACS), MAX-MIN Ant System, and others were produced [6]. By accurate definition of the problem, ACO can be useful in various areas.

4 An ACO Based Image Partitional Classifier

As stated before, the K-means algorithm tends to find the local minimum rather than the global because it is heavily influenced by the choice of initial centroids. The integration of ACO with a K-Means is able to omit undesired solutions and makes classification process less dependent on the initial parameters. A conventional clustering algorithms partition a data set into a set of clusters S_j, $j = 1,...,K$. A widely adopted definition of optimal clustering is a partitioning that the intra cluster similarity is minimized while the inter cluster similarity is maximized. For a given problem the pheromone can be set to be proportional to above criteria of the desired solution.

In our proposal, the ACO plays its part in assigning each image to a cluster and each ant is giving its own classification solution. The algorithm starts by assigning a pheromone level τ and a heuristic information η to each image. The value for the pheromone level τ assigned to each image is initialized to 1 so that it does not have effect on the probability at the beginning. Heuristic information $\eta_{(X_i,C_j)}$ is obtained from the following formula:

$$\eta_{(X_i,C_j)} = \frac{B}{Sim(X_i,C_j)} , \tag{6}$$

where X_i represents the feature vector of ith image and C_j is the feature vector representing j^{th} centroid of the cluster. $Sim(X_i,C_j)$ is the similarity between an image X_i and C_j. The constant B is used to balance the value of η with τ. Then each ant will assign each ith image to a jth cluster with the probability $P_{(X_i,C_j)}$ obtained from:

$$P_{(X_i,C_j)} = \frac{\tau_{(X_i,C_j)}\eta_{(X_i,C_j)}}{\sum_{j=1}^{K} \tau_{(X_i,C_j)}\eta_{(X_i,C_j)}} , \tag{7}$$

where K is the number of clusters. Assume a number m of ants is chosen for clustering based on the K-Means approach. After all ants have done their classification, the assigned pheromone to this solution is incremented. In order to find global minimum, the pheromone value is updated according to quality of the solution. For updating the pheromone to each clustering the following formula is used:

$$\tau_{(X_i,C_j)}(t) = \rho \cdot \tau_{(X_i,C_j)}(t-1) + \sum_{a=1}^{m} \Delta\tau_{(X_i,C_j)}{}^{a}(t) , \tag{8}$$

where ρ is the pheromone trail evaporation coefficient $(0 \le \rho \le 1)$ which causes vanishing of the pheromones over the iterations. $\tau_{(X_i,C_j)}(t-1)$ represents the pheromone value from previous iteration. $\Delta\tau_{(X_i,C_j)}{}^{a}(t)$ in Eq. 9 is a new amount of pheromones calculated from all m ants that assign image X_i to j^{th} cluster. This approach of marking solutions by pheromone is proposed as follows:

$$\Delta\tau_{(X_i,C_j)}{}^{a} = \begin{cases} \dfrac{N \cdot InterSim\,(a,j)}{IntraSim\,(a)} & \text{,if } X_i \text{ belongs to } j^{th}\text{cluster} \\ 0 & \text{, otherwise} \end{cases} . \tag{9}$$

$$InterSim(a, j) = \sum_{n \ne j}^{K} Sim(C_j, C_n) , \tag{10}$$

$$IntraSim(a) = \sum_{i \in S_j}^{N_j} \sum_{j=1}^{K} Sim(X_i, C_j) , \tag{11}$$

InterSim(a, j) represents the sum of the similarities obtained by ant a, between j^{th} centroid and the rest of centroids. *IntraSim*(a) represents the sum of the similarities obtained by ant a, between each image and its centroid. N is the number of images in the dataset and this variable keeps the values of *InterSim*(a, j) and *IntraSim*(a) in the same order. It becomes clear that the pheromone increases when clusters get more apart and when each cluster has more similar images. Next, the classification performed by each ant is driven by the quality of previous solutions. This is repeated until the best solution for all ants is achieved.

5 Experimental Results

In order to verify the performance of our method, some computer simulations have been conducted. In the simulation the parameters involved are set up as follows. Parameters B and ρ are set to be, $B = 100$ and $\rho = 0.2$. The number of ants is chosen to be $m = 100$. For feature representing of images was used MPEG7 - Color Layout Descriptor (CLD) with 58 coefficients. For matching two CLDs the following similarity measure was used:

$$Sim(X_i, X_j) = \sqrt{\sum_{k=1}^{28}(X_i(k) - X_j(k))^2} + \sqrt{\sum_{k=29}^{43}(X_i(k) - X_j(k))^2} + \sqrt{\sum_{k=44}^{58}(X_i(k) - X_j(k))^2} \quad (12)$$

where first 28 coefficients are Yi, next 25 Cb and last 25 Cr [5].

Experiments on binary classification of two different image databases are presented. "Window on the UK 2000" CD, available at http://www.bnsc.org/wouk/wouk1.htm, and consist of 390 images divided to 6 sets of similar blocks (Boats, Fields, Vehicles, Trees, Buildings, Roads). Second database was obtained from The Corel image database includes 700 images divided to 7 categories (Lions, Rural, Buildings, Cars, Elephants, Clouds and Tigers), all consist of 100 images. Representative samples of images from both databases are depicted in Figure 2.

The results for both the ACO-based Classifier and simple K-means are compared on various sets of images. Different runs of algorithms on both databases are shown in tables 1 and 2. In both cases the number of clusters K was set to be 2. Random classes of images were chosen in order to show classification results on visually similar or dissimilar input data. Results given by different runs of the K-Means algorithm were different due to randomly initial assigning of centroids. Therefore the K-Means algorithm was run 10 times and average precision and recall of solutions was computed. The ACO makes classification process stable hence multiple iterations were not needed. According to the classification comparisons from Table 1 and 2, it can be observed that the ACO based classifier has better results with respect to simple K-Means. Also, the above results show higher efficiency of the ACO in case of more similar sets of images. In the latter case, the initial assignment of cluster centroids

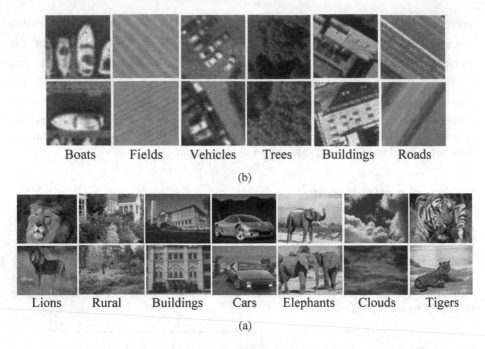

Boats   Fields   Vehicles   Trees   Buildings   Roads

(b)

Lions   Rural   Buildings   Cars   Elephants   Clouds   Tigers

(a)

Fig. 2. Several representative images for different categories taken from a) the "Window on the UK 2000" database, b) the Corel image database

visibly influences the classification process and decrease accuracy of the classification. On the contrary, ACO does not affect classification process on visually dissimilar images (e.g. building/field, fields/vehicles) when K-Means becomes more stable.

Table 1. Image Classification results on the sets of images obtained from the "Window on the UK 2000" database

Class1	Class2	K-Means			ACO-Classifier		
		Precision [%]	Recall [%]	Accuracy [%]	Precision [%]	Recall [%]	Accuracy [%]
Boat	Field	**94**	60	94	**94**	**64**	**96**
Boat	Vehicle	68	65	62	**73**	**70**	**70**
Tree	Building	76	67	85	**98**	**69**	**90**
Tree	Boat	**78**	49	71	77	**56**	**75**
Building	Field	**95**	**56**	**83**	95	56	83
Building	Road	60	45	60	**78**	**50**	**71**
Field	Vehicle	87	**68**	93	**90**	68	**94**
Field	Tree	98	60	96	**100**	**61**	**97**
Road	Vehicle	68	52	73	**78**	**57**	**79**
Road	Tree	88	54	82	**96**	**57**	**90**

Table 2. Image Classification results on the sets of images obtained from the Corel image database

Class1	Class2	K-Means			ACO-Classifier		
		Precision [%]	Recall [%]	Accuracy [%]	Precision [%]	Recall [%]	Accuracy [%]
Rural	Building	69	91	75	72	93	78.5
Lion	Car	60	52	62	68	56	72
Elephant	Cloud	82	93	85	84	95	88.5
Lion	Rural	57	68	59	65	72	66
Building	Car	70	94	76.5	75	94	79.5
Tiger	Cloud	79	73	82.5	77	76	84

6 Conclusion and Future Work

In this paper, we propose the ACO-based Classification algorithm to solve images classification problem. The proposed classifier combines the K-Means and biologically inspired learning mechanism based on pheromone-driven communication of ants. The K-Means is heavily influenced by the choice of initial cluster centroids and the distribution of data. The above experiments show the potential of using ACO to optimize the classification process. The ACO makes the K-means algorithm less dependent on the initial parameters; hence it makes it more stable. The combination of several MPEG7 descriptors and implementation of learning mechanism, therefore increasing the image classification accuracy will be part of future studies.

References

1. Colorni, A., Dorigo, M., Maniezzo, V.: Distributed optimization by ant colonies, In Proceedings of ECAL'91 European Conference on Artificial Life, Elsevier Publishing, Amsterdam, The Netherlands (1991) 134-142
2. Deneubourg, J., L., Aron, S., Goss, S. and Pasteels J., M.: The self-organizing exploratory pattern of the argentine ant. Journal of Insect Behavior, (1990) 159–168
3. Dorigo, M.: Optimization, Learning and Natural Algorithm. PhD thesis, Dipartimento di Elettronica e Informazione, Politecnico di Milano, IT, (1992)
4. Dorigo, M., Maniezzo, V. and Colorni, A. Positive feedback as a search strategy. Technical Report, Dipartimento di Elettronica, Politecnico di Milano, IT, (1991) 91-016
5. Manjunath, B.S., Salembiar, P., Sikora, T.: "Introduction to MPEG – 7, Multimedia Content Description Interface", John Wiley, (2003)
6. Merloti P. E.: Optimization Algorithms Inspired by Biological Ants and Swarm Behavior. Cabo Bahia, Chula Vista, CA 91914 United States of America, (2004)
7. Ramos, V., Almeida, V.: Artificial Ant Colonies in Digital Image Habitats - A Mass Behavior Effect Study on Pattern Recognition. Proceedings of ANTS'2000 - 2nd International Workshop on Ant Algorithms (From Ant Colonies to Artificial Ants), Brussels, Belgium, 7-9, (2000) 113-116

8. Saatchi S., Hung Ch. Ch.: Hybridization of the Ant Colony Optimization with the K-Means Algorithm for Clustering, SCIA 2005, Springer-Verlag Berlin Heidelberg (2005) 511 – 520
9. Xu R. and Wunch D.: Survey of Clustering Algorithms, IEEE Trans. Neural Network, Vol.6, No.3 (2005) 645-678

Use of Image Regions in Context-Adaptive Image Classification*

Ville Viitaniemi and Jorma Laaksonen

Laboratory of Computer and Information Science, Helsinki University of Technology,
P.O.Box 5400, FIN-02015 TKK, Finland
{ville.viitaniemi, jorma.laaksonen}@tkk.fi

Abstract. In this paper we describe and discuss our existing PicSOM software framework from the point of view of context-adaptive analysis of image contents, especially its method for using automatic image segmentation. We describe and experimentally validate a modification to the segment-using procedure that both essentially reduces the computational cost and slightly improves classification accuracy. Finally, we apply the segment-using methodology in qualitatively investigating the roles of primary objects and their context in classifying the images of the Pascal VOC Challenge 2006 database.

1 Introduction

People are nowadays faced with constant and overwhelming stream of digital image and video content. Furthermore, the amounts of data being generated seems to be constantly increasing. Therefore, automatic methods are highly desirable to analyse and index the large data masses. Especially useful would be methods that could automatically interpret the semantic contents of images and videos as it is just the content that determines the usefulness of them for most purposes.

Our PicSOM software framework [8,11] is aimed at automatically organising and ordering large unannotated databases of audio-visual objects according to the similarity of their contents. The databases may include images, videos, texts and multi-part objects, such as mobile multimedia messages, emails and web pages. The objects in the database may be hierarchically related to each other, such as different parts of a multimedia message. In this paper, we focus on the case of image databases where the hierarchy consists of images and their segments. The underlying principle of the framework is to extract a large number of visual features from the database images. Then the aim is to find statistical dependencies between the visual features and the current goals of image analysis. The dependencies are not required to be deterministic rules, such as blue always

---

* Supported by the Academy of Finland in the projects *Neural methods in information retrieval based on automatic content analysis and relevance feedback* and *Finnish Centre of Excellence in Adaptive Informatics Research*.

Y. Avrithis et al. (Eds.): SAMT 2006, LNCS 4306, pp. 169–183, 2006.
© Springer-Verlag Berlin Heidelberg 2006

corresponding to sky. On the contrary, even weak probabilistic correlations are sought after.

A natural approach to describing an image would be to list what different parts the image contains, and then possibly describe the major parts in more detail. In this light, partitioning the image to disparate parts and describing the image in terms of the content and relationship of these parts appears to be a promising approach. Indeed, image segmentation often is crucial in image understanding. The PicSOM method of using image segments is in line with the statistical approach to image analysis: a set of alternative segmentations is formed and the system is adaptively allowed to find the segmentations most beneficial to the image analysis task at hand.

On different occasions, the similarity of image content arises from different visual features, such as colour or shapes. Only a subset of all the imaginable features is relevant in a given image analysis task. It is the context of image analysis that determines the set of relevant visual features. For purposes of the PicSOM framework, we divide the context into two levels of specificity. The distinction is made to allow two different mechanisms of adaptation of the framework to the context. We call the broader level of context database context. With database we understand the collection of the images the content analysis is targeted at, whether or not they are actually collected in a database. We consider the database part of the context to be fixed, so the system may be adapted to the database level context beforehand, without needing to take the suitability of this specific system to other databases. A schematic example of database level adaptation to context is given by an imaginary database consisting solely of red and green apples. In the context of this database, it is not wise to allocate resources to a colour feature telling whether an object is blue. Task level context is the part of the context that changes so frequently that completely re-building the image analysis system would be impractical. For example, we may think of a database of vehicles of different colour. Different visual features are relevant, depending on whether we want to identify motorbikes, red automobiles, or any yellow vehicle in the database.

In this paper we consider the use of the adaptive PicSOM image analysis framework in the context of image classification task detailed in Section 2. Section 3 outlines the working principles of the framework in general, whereas in Section 4 we discuss the use of image segments more in detail and experimentally investigate of the usefulness in the image classification task. In Section 5 we describe how the method for using segments may be moved from the on-line part of the system to the off-line preprocessing and made computationally more lightweight.

Having performed the image classification task with help of image segments, we may reverse the direction of the analysis. In Section 6 we compare the the contribution of different image segments to the classification in the context of this image database. In Section 7 we present conclusions and discussion.

2 Image Analysis Task

To evaluate and motivate the techniques we discuss later in this paper, we consider a concrete image analysis task. The task consists of classifying images of the portion of the Pascal Visual Object Classes Challenge 2006[1] image set whose ground truth classifications have been made public at the time of this writing. This set of 2618 images contains realistic images of ten classes. The classes are defined by the absence or presence of an object, e.g. cat, in the images. The classes are partially overlapping, i.e. several different objects may appear in the same image.

The classification task is considered in a supervised setting: approximately one half of the images is used as training set and the rest as the test set. Table 1 shows the class statistics of the image sets. The classification performance is evaluated in terms of Area Under Curve (AUC) property of the classifier Receiver Operating Characteristic (ROC) curves.

Table 1. Statistics of image sets. Columns correspond to different object classes.

	bicycle	bus	car	cat	cow	dog	horse	motorbike	person	sheep	total
training set	127	93	271	192	102	189	129	118	319	119	1277
test set	143	81	282	194	104	176	118	117	347	132	1341

3 PicSOM Image Analysis Framework

3.1 Outline

In the PicSOM framework the database objects are divided into two groups: example and target objects. In a supervised learning context these correspond to training and test sets, respectively. In an interactive image retrieval setting the set of example and target objects is adapted dynamically during a retrieval session as a result of the user giving relevance feedback on the target objects the system retrieves. The framework is used to rank the target objects according to their similarity to a given set of positive example objects, simultaneously combined with the dissimilarity to a set of negative example objects.

The similarity of the objects is evaluated in terms of a large set of visual features of statistical nature. For this purpose the example and target images are pre-processed in the same manner: the images are first automatically segmented and a large collection of statistical features is extracted from both the segments and whole images. Several different segmentations of the same images can be used in parallel. Procedures for feature extraction and image segmentation are discussed in detail in Sections 3.2 and 4, respectively. In the current experiments, we consider the use of 290 segmentation–feature combinations.

---

[1] http://www.pascal-network.org/challenges/VOC/

The features extracted from images and image segments can be interpreted as feature vectors in multi-dimensional feature spaces. Each of the resulting feature spaces is quantised with a tree-structured variant of the Self-Organising Map (SOM) [6], a TS-SOM [7]. SOM is an unsupervised neural algorithm that adaptively forms a mapping from a high-dimensional input space onto a two dimensional grid. The database level adaptation to context in the PicSOM framework relies just on the adaptiveness of the SOM: the quantisation of the feature spaces concentrates attention on feature values that are actually common in the database.

The quantisation forms representations of the feature spaces where points on the two-dimensional TS-SOM surfaces correspond to images and image segments. For the current experiments, 64x64 TS-SOMs are used. The organisation of the image database produced by one of the feature TS-SOMs is shown in Figures 1 and 2.

Due to the topology preserving property of the TS-SOM mapping, assessment of image similarity in terms of each of the individual feature spaces can be performed by evaluating the distance of the representation of an object on the TS-SOM grid to the representations of positive and negative example objects. Technically this is done by placing triangular convolution kernels to the locations of example objects and evaluating their value at the locations of the target objects. The kernels placed at positive examples are normalised to sum to unity, as are the kernels placed at negative examples.

Fig. 1. A TS-SOM organised by colour moments of image patches. For sake of clarity, a 16x16 intermediate level of the TS-SOM structure is shown. For actual image analysis, only the 64x64 bottom level is used.

Fig. 2. A close-up of the bottom left corner of the colour moment TS-SOM that represents image patches with red colour together with light-coloured structures

A task-level adaptation mechanism of the PicSOM framework is provided by the way the similarities in different feature spaces are combined. Due to the performed normalisations, such feature spaces are effectively emphasised that perform best in discriminating the positive and negative example objects when the similarities are summed together. This is because on poorly performing feature TS-SOMs, the positive and negative kernels intermingle on the same map areas and effectively neutralise each other. On the other hand, on well discriminating TS-SOMs, the positive and negative kernels concentrate on separate areas where they amplify each other. This results in large amplitudes of similarity and dissimilarity peaks. Because the convolution kernels have a limited and fixed support, the similarity assessment procedure only takes into account local distances. This is also desired as the TS-SOM mapping only preserves local topology.

Often the goal of the system is to rank images, not just image segments according to their similarity. Segment-level similarity must thus be combined into image-level similarity. The solution used in PicSOM is straightforward: image-level similarity is obtained by summing the contributions of the segments of the image. This technique is further discussed, evaluated and extended in Sections 4 and 5.

3.2   Visual Features

A number of statistical visual features is extracted and made available for the similarity assessment algorithm. The features include MPEG-7 standard descriptors [4] as well as some non-standard descriptors. The features are extracted from image segments as well as from whole images when appropriate. Table 2 lists the used visual features.

The MPEG-7 features are extracted using the MPEG-7 Experimentation Model. For some image segments, we have replaced the MPEG-7 Color Layout and Scalable Color descriptors with our own approximate implementation of the standard for computational reasons.

In addition to the tabulated features, we have found that composite features formed from pairs of visual features are highly beneficial in our system. Currently

Table 2. Visual features extracted from image segments

MPEG-7 descriptors	non-standard descriptors
Color Layout	average colour in CIE L*a*b* colour space
Dominant Color	central moments of colour distribution
Region Shape	Fourier descriptors of segment contours
Scalable Color	histogram of Sobel edge directions
	co-occurence matrix of Sobel edge directions
	Fourier transform of local edges
	8-neighbourhood binarised intensity texture
	Zernike moments of binarised segment shape

the composite features are formed by simply concatenating the corresponding feature vectors and equalising the variance of the vector elements. We have confined us to pairs of features as the number of larger feature combinations grows very rapidly. For the same reason, we have not formed all the possible pairs but picked some most promising combinations.

4 Image Segmentation

Currently there appears to be no definitive knowledge concerning what sorts of image segmentation is beneficial or adequate for content-based image analysis. Generic, complete and to-the-pixel accurate unsupervised segmentation is generally regarded to be virtually impossible. Even if it was possible, one still has to address the issue of diversity of segmentations: one image can typically be segmented in numerous different perfectly valid ways. Choosing one segmentation over another would require interpreting the image already in an early stage of the image processing chain. A subset of the diverse segmentations of an image is formed by hierarchically related segmentations: the objects in the images often naturally decompose into sub-objects, thus defining object-part hierarchy trees.

4.1 Segmentation in PicSOM Framework

The approach PicSOM uses for image analysis is statistical in nature. In the same spirit, PicSOM also uses image segmentation in statistical manner. A number of uncertain segmentations is generated by simple means. Every single segmentation may not be correct or visually viable, but on the statistical level we hope to observe correlations between the visual properties of the segments and the sets of example images.

Besides perfect segmentations being impossible to obtain in practice, it still remains to be demonstrated how much the performance of image content analysis is compromised if less complete segmentations are used. Of course, several factors affect this question. One of them is the sophistication level of the image analysis approach. It is reasonable to suppose that if segmentations are used in a statistical manner, fairly inaccurate image segmentation may be enough.

Also the different visual features set different needs for image segmentation. For example, colour and texture features are quite tolerant to inaccuracies in segmentation, whereas some shape features may critically depend on the quality of segmentations.

It seems natural to think that more deterministic image content analysis could lead to better results. Then there would also be need for more accurate image segmentations. However, the more accurate associations would likely be specific to specific classes of example images, such as cars. On the other hand, we want to keep our framework generic, independent of specific types of images. The framework thus has to autonomously learn the associations between image classes and their visual properties. At the current state of machine learning, learning of much more sophisticated rules seems difficult. Certainly we aim at pushing the border into that direction, but currently much more sophisticated associations would probably have to be manually specified to the system.

Ideally, the division between image segmentation and feature extraction system stages would be clear-cut. In practice, however, this is not always the case. Many visual features codify also spatial information: example of this is the MPEG-7 standard Edge Histogram feature that implicitly divides the image into 16 fixed tiles.

4.2   Implemented Segmentation Methods

The actual segmentation methods that are used to generate the alternative segmentations of the images are rather rudimentary. One of the methods geometrically divides the images to overlapping squares. Side length of the square is determined by dividing the larger of the image dimensions by ten. The overlap is achieved by first non-overlappingly tiling the image and then placing another set of tiles at the crossings of the tiling. This results in no more than 181 (for square images) square segments, on the average about 140.

As another segmentation method we employ an area-based region merging algorithm based on homogeneity in terms of colour and texture. The merging is continued until a fixed number of image segments are left. Two sets of segmentations is obtained with two different sets of region merging parameters. The first parameter values give rise to 8 segments whereas the lattar result in 25 segments. The parameters for the merging algorithm have been selected to give visually feasible results for photographs and other images in earlier applications. The diversity of segmentations is increased by recording, in addition to basic segments, the hierarchical segmentation that results from continuing the region-merging algorithm until only three regions remain. Altogether we thus have five segmentations of each database image.

4.3   Combining Segment Similarity as Image Similarity

As briefly mentioned in Section 3, in PicSOM framework the segment-level similarities are just summed together to form the image level similarities. Re-ordering summations, the similarity comparison procedure in a single feature space i can be recast as

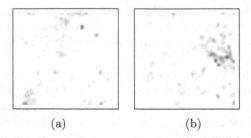

(a)   (b)

Fig. 3. Smoothed histograms of positive example segments in the quantised colour moment feature space: (a) image class "bus", (b) image class "cow". Dark shades correspond to large values.

1. Accumulating the normalised two-dimensional histograms E_i^+, E_i^- and T of positive example segments, negative example segments and target segments, respectively, using the i:th TS-SOM to define the histogram bins.
2. Kernel smoothing the subtracted histogram of example segments: $H_i = K * (E_i^+ - E_i^-)$, where K is a two-dimensional triangular kernel.
3. Interpreting the histograms as vectors and calculating the similarity as inner product: $S_i = < T_i, H_i >$.

Figure 3 shows two smoothed histograms of positive example segments in the colour moment feature space quantised with the TS-SOM of Figure 1. Among the background noise and spurious responses, we can identify several plausible concentrations from the histograms. In the left subfigure, the distribution of image class "bus" concentrates especially in the lower left corner and upper border of the SOM surface. These areas correspond to image patches with red structures. This is explained by the fact that majority of the buses are British red buses. The segments of image class "cow" concentrate on the rightmost part of the SOM that represents green image patches and image patches with structures on a green background.

Histograms summarise the probability distributions strongly: they introduce large number of invariances. Additional sources of invariances, i.e. information loss, in the present use of image segments are the complete ignorance on the spatial distribution of feature values and failure to connect the same segment in different feature spaces. Example of former would be to consider blue patches of sky in the upper part of an image to be equivalent to the blue patch of water in the lower part. Example of the latter would be to consider images with blue ball and red triangle equivalent with an image of a blue triangle with a red ball.

In some approaches [14,2,5] example and target image segments are explicitly matched with each other. With the numerous alternative segmentations and feature spaces PicSOM uses, the combinatorics of such approach soon become prohibitive. The introduced invariances, however, make the learning problem much less complex and thus more manageable. Learning with less invariances requires more learning material, i.e. example images. The resulting computational costs could easily rise to currently unrealisable level.

5   Off-Line Procedure for Using Image Segments

In the previous section, we described a method of using image segments that has been found to be useful in many image tasks. The use of image segments introduces an increased computational cost, however. For some on-line purposes such as interactive image retrieval, large time complexity may not be acceptable. Therefore, we have devised a method for precomputing a representation of the segment contents off-line as part of adaptation to the database level context and just comparing the representations on-line. For an image and a given feature space, the representations are obtained as follows:

1. A two-dimensional histogram of the feature values of the image segments are accumulated.
2. The histogram is kernel smoothed with a triangular kernel K.
3. The histogram with is downsampled with a factor of four.
4. The histograms are regarded as vectors and their dimensionality is reduced to a fixed number with principal component analysis (PCA).
5. normalise the vector components to unit variance.

For combining multiple feature spaces, we concatenate the histograms as the vectors after downsampling in step 2 and perform PCA and normalising just as for single feature spaces. Compared to only PCA, additional normalising of the components was observed to enhance image classification performance. Further processing by estimating the linear ICA model [3] from the vectors did not bring improvements in connection with the classification procedure used in the experiments.

There are two main parameters of the method: the width w of the smoothing kernel K and the number of components l_{PCA} left after the PCA. A series of experiments were performed to determine optimal values for the parameters. Unfortunately, the best values seem to be feature and image class specific. In determining a suitable compromise we decided to put more emphasis on features and image classes where the descriptor performed well in the classification task. The rationale for this is that probably the poorly performing features will not be used, after all, for that specific image class as there is a wide variety of other features available. In our experiments, we have ended up in parameter values $w = 24$ for the convolution kernel width and $l_{PCA} = 64$ for the dimensionality of the PCA.

For the initial verification of the usefulness of the obtained region representations, the image classification performance resulting from their use was evaluated using a support vector machine implementation as a black-box classifier. As SVM, we use the C-SVM implementation with RBF kernels implemented in the libsvm software package [1]. With this implementation, the probability estimates needed for ROC curve formation are obtained by the pairwise coupling method described in [15].

5.1   Qualitative Validation

We have compared the image classification performance of the introduced method for obtaining the obtained off-line image representations against the traditional PicSOM on-line algorithm. For this initial experiments, we use a segmentation–feature combinations that were conveniently available. As image segmentation method we employ the square-tile-segmentation described in Section 4.2. We consider the image classification performance of the image representation derived from both 1) a single feature: color layout, and 2) a combination of three features: color layout, Fourier transform of edge distribution and color histogram/texture descriptor, the last feature being a concatenation of two features. It turns out that the choice of the specific features is somewhat unfortunate, as their performance in classifying these image classes is not very good and combining the features does not help the on-line PicSOM algorithm. However, the relative differences between the on-line and off-line variants can still be observed. Figures 4 and 5 visualise the feature space corresponding to the combined off-line representation by means of a SOM. We see that in general, some of the images cluster nicely on the SOM, for example the cluster of cars near the bottom right corner, whereas there are also some areas where images of various objects mix.

5.2   Quantitative Validation

Figure 6 compares the image classification performances of the described off-line and on-line methods for combining image segments. With the single feature, the relative order of the methods varies, the off-line variant perhaps being slightly

Fig. 4. An intermediate 16x16 level of a TS-SOM visualisation of the feature space defined by the offline image representation derived from the combination of three image segment features

Fig. 5. A close-up of area near the bottom right corner corner of the TS-SOM visualisation of Figure 4

better overall. With the combined feature, the off-line method is consistently better than the on-line method with the exception of the accuracy being approximately equal for one image class. It is notable that the off-line method always manages to benefit from the incorporation of additional features, which is not always the case with the on-line method.

Besides the classification accuracy, an important factor in choosing between the algorithmic variant is speed. After precalculating the off-line representations, the actual comparison of the introduced region representations is observed to be more than 1300 times faster than executing the on-line algorithm. On the other hand, the on-line algorithm is more flexible. It allows the set of positive example segments to be specified freely on run-time, whereas the offline representation summarises all the segments in an image. The flexibility might be needed, for example if the framework is used for image analysis on the level of image segments. This is the case in [13] where the framework is discriminatively used to locate an object in the images an basis of examples, in [12] where image-level keywords are focused on specific image segments, and in the next section of this paper.

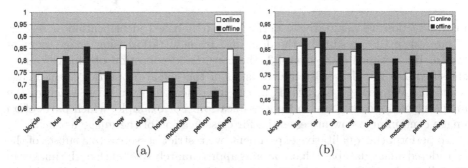

Fig. 6. AUC performance of image classification of both on-line (white bars) and offline (black bars) methods for combining image segments for the ten VOC object classes. The two subfigures shows the accuracy resulting from use of (a) a single colour layout feature, and (b) the combination of three features.

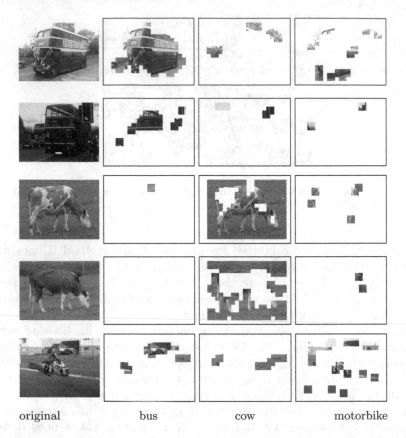

original   bus   cow   motorbike

Fig. 7. Images of the image collection shown together with the collection-wide top segments that contribute most to the classification of the images as "bus", "cow" and "motorbike". The original images are shown in the leftmost column.

6   Context-Based Identification of Regions of Interest

In this section we qualitatively perform context-based identification of regions of interest, made possible by the described framework for using image segmentation. By methods of Section 4.3, the considered collection of images can be classified into ten classes by the statistics of the visual properties of the image segments. We now go backwards in the inference chain and ask which of the image segments contribute to the classification. For example: which parts of an image containing a cow are essential for classification of the image as such?

To perform this qualitative experiment, we restrict ourselves to a subset of the considered image database that contains approximately one tenth of the images in the full database. The class composition remains approximately equal. For the experiments we use the geometrical square tile segmentation described in Section 4.2.

Figure 7 has been generated by marking to the images the segments that contribute most to the classification of images. The threshold has been set globally to include one tenth of segments of the image database. In the figure each row corresponds to a separate image. The original image is shown in the leftmost column and the remaining columns display such segments of the image that contribute to the image's classification as "bus", "cow" and "motorbike", respectively. In Figure 8 we show a more comprehensive set of similar illustrations of image segments that contribute to classification as "bus".

Fig. 8. The locations of image-collection-wide top bus segments in some images actually depicting buses

In an ideal case, in the figures all the image segments containing the class defining object would be highlighted, whereas all the segments outside these objects would remain unhighlighted. Then we would have directly obtained a context-based method for segmenting images contained in an image database. The method would be adaptive to the image database in two senses. Firstly, the context of the database would determine the visual elements that are discriminative in that certain database. Secondly, the segmentation method would be class-specific, producing segmentations that are discriminative in the sense of the specific object class the segmentation method was derived from.

From the images we see, however, that the situation is far from the above described ideal case. Still, from Figure 7 we see that there is strong tendency of the algorithm to identify apparently reasonable parts of the images as responsible for the classification. The algorithm is also class specific: for instance only few segments of the bus images would be likely to appear in the cow images and vice versa. This property is not deterministic, however, but probabilistic in nature.

In many cases, the context of the objects is found to be equally important as the visual properties of the object themselves. This can be seen for example in Figure 8 where the surrounding traffic or road scene often is included in the most contributing parts of the bus images. Another observation is that inside an object only some of its parts may be specific to that object class. This is exemplified, for example, by the motorbike image in Figure 7. The cow images also exemplify well these two phenomena. Both can also be explained by the limited scope of the underlying image database. The context of this type of green grass seems to be specific for the cow images in this database, although the database contains other animals such as sheep and horses. Apparently the grass in those images is somehow different. On the other hand, the brown fur part of the lower cow image does not contribute much to classification as cow. This can be explained by the database containing also other animals such as horses and dogs. In contrast, the contours of lower part of the cow, especially legs seem to be characteristic for cows.

In this section we have seen that using the above described image representation in terms of regions, we can back-track in the image classification algorithm and identify plausible regions of interest. However, the connection between the identified regions and genuine object segmentation is not straightforward. Still, integrated with other image segmentation, the regions of interest might provide a helpful cue in segmentation.

7   Discussion and Conclusions

In this paper, we have discussed the method of using image segments in the Pic-SOM statistical image analysis framework. A procedure was devised for moving the computational demands of this segment-using method from on-line computation to the off-line phase, thus reducing the on-line complexity to a small fraction. Still, the image classification accuracy remained uncompromised in the experiments performed. The obtained time savings are important for keeping the computational demands of the task-level context adaptation manageable. The performed experiments were initial and must be complemented with further testing with a more comprehensive set of visual features and larger image databases.

We saw that the current method of using segments introduces many invariances. Although lifting the invariances could easily lead to need for huge amounts of example data, we are still planning to consider lifting some of the invariances, for example by considering spatial relationships of the segments [9]. The beneficiality of geometric constraints to the accuracy of image analysis has been demonstrated e.g. in [10].

We can interpret the method of generating the off-line representations of the segment distributions as a novel method of generating image features adapted to the current image analysis context, given the segment level visual features and the image segmentations. The method adapts to the database level context through the adaptive feature representations formed using SOMs. Also the segmentations may be adaptive, either on the database level or on the task level, as outlined—although not implemented—in Section 6. Comparing the merits of these adaptive features to customary, non-adaptive image features such as the MPEG-7 features remains to be performed in the future.

We qualitatively investigated the contribution of different types of segments to image classification in Section 6. It was observed that in the context of the current VOC image database, contextual information was often as important for the classification as the target object itself. This result is naturally dependent on the database at hand. If the database is sufficiently comprehensive, the significance of context in the images may become lesser, and the actual target object in the image becomes the most important contributor in image classification, as in [12].

References

1. Chih-Chung Chang and Chih-Jen Lin. *LIBSVM: a library for support vector machines*, 2001. Software available at http://www.csie.ntu.edu.tw/~cjlin/libsvm.
2. Yixin Chen and James Z. Wang. Looking beyond region boundaries: Region-based image retrieval using fuzzy feature matching. In *Multimedia Content-Based Indexing and Retrieval Workshop, September 24-25*, INRIA Rocquencourt, France, September 2001.
3. Aapo Hyvärinen, Juha Karhunen, and Erkki Oja. *Independent Component Analysis*. John Wiley & Sons, 2001.
4. ISO/IEC. Information technology - Multimedia content description interface - Part 3: Visual, 2002. 15938-3:2002(E).
5. F. Jing, M. Li, L. Zhang, H. Zhang, and B. Zhang. Learning in region-based image retrieval. In *Proceedings of International Conference on Image and Video Retrieval*, volume 2728 of *Lecture Notes in Computer Science*, pages 198–207. Springer, 2003.
6. Teuvo Kohonen. *Self-Organizing Maps*, volume 30 of *Springer Series in Information Sciences*. Springer-Verlag, Berlin, third edition, 2001.
7. Pasi Koikkalainen. Progress with the tree-structured self-organizing map. In *11th European Conference on Artificial Intelligence*. European Committee for Artificial Intelligence (ECCAI), August 1994.
8. Jorma Laaksonen, Markus Koskela, and Erkki Oja. PicSOM—Self-organizing image retrieval with MPEG-7 content descriptions. *IEEE Transactions on Neural Networks*, 13(4):841–853, July 2002.
9. Jorma Laaksonen and Ville Viitaniemi. Emergence of ontological relations from visual data with self-organizing maps. In *Proceedings of the 9th Scandinavian Conference on Artificial Intelligence Scandinavian*, Espoo, Finland, October 2006. To appear.
10. C. Millet, I. Bloch, P. Hède, and P. A. Moëllic. Using relative spatial relationships to improve individual region recognition. In *Proceedings of 2nd European Workshop on the Integration of Knowledge, Semantic and Digital Media Technologies*, pages 119–126, London, UK, November 2005.
11. Mats Sjöberg, Jorma Laaksonen, and Ville Viitaniemi. Using image segments in PicSOM CBIR system. In *Proceedings of 13th Scandinavian Conference on Image Analysis (SCIA 2003)*, pages 1106–1113, Halmstad, Sweden, June/July 2003.
12. Ville Viitaniemi and Jorma Laaksonen. *Focusing Keywords to Automatically Extracted Image Segments Using Self-Organising Maps*, volume 210 of *Studies in Fuzziness and Soft Computing*. Springer Verlag, 2006. To appear.
13. Ville Viitaniemi and Jorma Laaksonen. Techniques for still image scene classification and object detection. In *Proceedings of 16th International Conference on Artificial Neural Networks (ICANN 2006)*, September 2006. To appear.
14. James Z. Wang, Jia Li, and Gio Wiederhold. SIMPLIcity: Semantics-sensitive integrated matching for picture libraries. *IEEE Transactions on Pattern Analysis and Machine Intelligence*, 23(9):947–963, September 2001.
15. T.-F. Wu, C.-J. Lin, and R.C.Weng. Probability estimates for multi-class classification by pairwise coupling. *Journal of Machine Learning Research*, 5:975–1005, 2005.

BPT Enhancement Based on Syntactic and Semantic Criteria*

C. Ferran, X. Giró, F. Marqués, and J.R. Casas

Signal Theory and Communications Department,
Technical University of Catalonia (UPC)
Campus Nord, C/ Jordi Girona, 1-3,
08034, Barcelona

Abstract. This paper presents two enhancements for the creation and analysis of Binary Partition Trees (BPTs). Firstly, the classic creation of BPT based on colour is expanded to include syntactic criteria derived from human perception. Secondly, a method to include semantic information in the BPT analysis is shown thanks to the definition of the BPT Semantic Neighborhood and the introduction of Semantic Trees. Both techniques aim at bridging the semantic gap between signal and semantics following a bottom-up and a top-down approach, respectively.

1 Introduction

1.1 Problem Statement: The Semantic Gap

Scene analysis for object extraction is a challenging and unsolved problem. The main difficulty of this task is to connect semantic entities to low-level features or vice-versa; that is, to fill the so-called **semantic gap**. There mainly exist two conceptually opposed approaches: **Top-down** methods, which are knowledge-based and search for pre-defined models into the image given some prior information. **Bottom-up** methods, which are generic methods aiming at linking visual features to perceptual meaningful primitives.

Moreover, methods based only on either top-down or bottom-up approaches might fail in a general context. On one hand, bottom-up methods cannot infer object entities without prior models. On the other hand, top-down techniques may benefit from a richer pre-analysis based on perceptual considerations. Thus, both approaches are not mutually exclusive, on the contrary, they might be coupled in order to succeed in the object extraction task. For this reason, a **common framework** is required.

As presented in [1], a **Binary-Partition-Tree** (BPT) is a structured representation of a set of hierarchical partitions usually obtained by means of an iterative segmentation procedure based on the optimization of an initial partition. The BPT is built up by iteratively merging the most similar pair of regions

---

* This work has been partly supported by the EU Network of Excellence MUSCLE, and by the grant TEC2004-01914 of the Spanish Government.

Y. Avrithis et al. (Eds.): SAMT 2006, LNCS 4306, pp. 184–198, 2006.
© Springer-Verlag Berlin Heidelberg 2006

given a homogeneity criterion. In this work the initial partition is assumed to include all the contours of the desired object. If not, one can always start from a *finer* segmentation. Therefore, the BPT representation can theoretically lead to a tree where each desired image object is represented by a single node at a certain level of the hierarchy. However, as pointed out in [2], from the bottom-up point of view there are some drawbacks related to the *simplicity* of the homogeneity properties used to construct the tree. This limitation can result in a tree where the desired object is not represented by a node. From the top-down point of view, not all the regions defined by the BPT are significant. Still the BPT representation offers the support for the extraction of objects and a common framework where regions and semantic entities can be linked. The goal of this paper is to overcome some of the previous limitations to enhance the BPT framework with the long term aim of bridging the semantic gap.

1.2   Proposal: Enhanced BPT as Common Framework

This paper presents the BPT as a common framework for bottom-up and top-down analysis. The framework has been improved from both points of view:

– **Bottom-up BPT construction:** Introducing and combining multiple and generic homogeneity criteria based on low- and middle-level features. Such features are refereed to as **syntactic** features, since they are defined by the relative positions of the regions they represent.
– **Top-down BPT analysis:** The problem of detections of a single instance of the same object is assessed. To do so, a model for semantic classes and its application on BPTs is presented.

The paper is organized as follows. Firstly, we present the motivations for both, the syntactic approach for BPT construction and the subsequent semantic BPT analysis. Secondly, we describe the enhanced BPT common framework and, some object detection and extraction examples are presented. Finally, the conclusions and the future worklines are drawn.

1.3   Motivations for the Syntactic Approach

When analysing an image, colour, motion or even depth information are usually prioritized disregarding other natural features of the objects belonging to the scene, such as symmetry or partial inclusion. Such features might be classified as being in the middle level vision problem.

This work relies on the assumption that, such complex features can be used to group image elements in order to form parts of objects or even complete objects. This assumption is based on the Gestalt psychology and on perceptual grouping approaches, as presented in [3] and [4], respectively.

In this context we assume that a **composite** object is an object which is built up of a set of distinct parts in various ways of **complex** composition. A well known technique for the study of such composition properties is syntactic image analysis [5]. Such work deals with the notions of primitive, grammar and syntax analysis from a formal point of view, ideas that have inspired the creation of the proposed syntactic framework for image segmentation.

This proposal is based on two key aspects. The definition and extraction of the syntactic features expressed as rules, and the critical decision of how to combine the different features. Regarding this second aspect, there may be situations in which the correct combination of different features (for instance, "similar colour as" and "partially included into") to create an object is not easy even for a human observer. In this work we attempt to provide a solution for such cases. Our proposal aims to apply the most significant rule given a set of features. This solution is based on the statistical analysis of the features over the whole image.

As a conclusion, we can define the syntactic image segmentation as an iterative region merging procedure based on the assessment of multiple homogeneity criteria expressed through rules which are combined according to their statistics. At each iteration the possible conflict between the defined rules is solved resulting in the merging of the most significant pair of regions.

1.4   Motivations for the Semantic Approach

Many analysis tasks focus on the extraction of semantic information from visual content. In applications such as object detection or content-based image retrieval, the final goal is to extract semantic entities from the visual data, as this is the most natural form for humans to access content. In most of the cases, the result of a segmentation and its organization in BPTs lead to many more regions than semantic entities are present in the content. Most of the regions represented by the BPT nodes lack a semantic meaning, although its organization in BPTs facilitates their analysis.

Partitions usually present oversegmentation problems, splitting a single object among more than a region. Nevertheless, this situation may be solved thanks to the hierarchical structure of the BPT. However, the BPT may contain a few nodes related to the same object which, in turn, may present similar perceptual properties and, for this reason, they may be considered as multiple instances of the same class if the analysis is based only on low-level features. Fig. 1 depicts an example of these cases, which represents how a fading rectangle is split in two during the segmentation process.

Selecting which of these similar regions represents the instance of a semantic class requires a previous semantic knowledge. This paper will propose a method to model semantic classes and how these models can help in a better semantic analysis of BPTs.

2   Enhanced BPT Framework

A BPT is created from an initial partition of an image, plus a set of hierarchichal partitions that are "'above" the initial partition. The finest level of detail is given by the initial partition (the leaves of the BPT). The nodes above the leaves

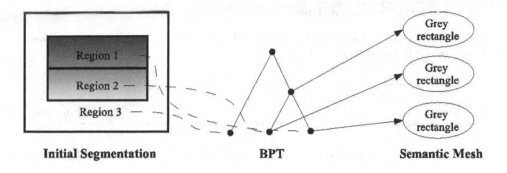

Fig. 1. Oversegmentation and BPT generates multiple instances of the same class

are associated to regions resulting from the merging of two children regions until we reach the root node, which represents the entire image support. An example illustrating the BPT for an image made of two rectangles over a white background can be seen on Fig.1.

2.1 Bottom-Up: Syntactic BPT Creation

The framework we present for BPT creation is an extension of the segmentation algorithm presented in [1], which performs an iterative region-merging. In this framework, an initial over-segmented partition is optimized by merging pairs of regions following a pre-defined optimization criterion. The BPT is built up by tracking the successive mergings of the algorithm. The aim is to iteratively estimate the region of support of homogeneous elements based on the features of the elements. The optimization criteria are based on the homogeneity of the elements and are defined as homogeneity criteria. For this purpose we define two type of descriptors:

1. Visual descriptors (VDs). These descriptors are computed locally, in a region neighboorhood, and provide individual dissimilarity measures for:
   Simple homogeneity criteria evaluated over pixel values within a single element and computed with *simple* operations between elements (pixels or regions). Colour, or region size are examples of descriptors computed using simple homogeneity criteria.
   Complex homogeneity criteria evaluated for two or more elements and computed using *complex* operations between elements (regions). Syntactic descriptors, such as symmetry or partial inclusion, are based on complex homogeneity criteria.
2. Statistical descriptors modelling the statistical distribution of the VDs and computed over the whole image, globally. For example, the entropy H_r of each rule r provides a measure of its significance in the current iteration. The statistical information allows to manage multiple homogeneity criteria by the use of global information.

188   C. Ferran et al.

2.2   Syntactic Segmentation

The major segmentation steps are simplification, feature extraction and decision, as presented in the scheme of Fig. 2.

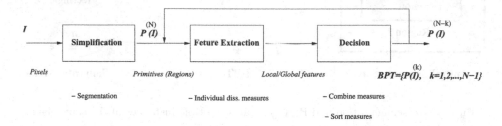

Fig. 2. Scheme of the major segmentation steps

Simplification. The purpose of the simplification step is to control the amount and nature of the information preserved for further analysis. Starting at the pixel level it generates a set of regions as primitives for the following steps. The resulting partition provides a robust information-rich set of region primitives overcoming the limitations of point-based primitives for higher level analysis.

Feature extraction. The feature space is created by associating a dissimilarity measure to each pair of regions given one of the following rules: R_i **has similar mean colour than** R_j, R_i **has similar size than** R_j **and is** R_i **partially included into** R_j. Moreover we impose the constraint that only neighbouring regions can be merged so that the result of any merging is still a partition of connected regions. The dissimilarity measure between regions R_i and R_j in the $k-th$ iteration using the $c-th$ rule is noted as $d_c^k(i,j)$. Its value tends to zero if the proposition is found to be true.

Even though the rules are computed locally over a pair of regions, an estimate of the distribution of the dissimilarity measures for each rule is computed by means of the histogram. In the case of colour homogeneity, the histogram counts the number of times that the same colour difference occurs among the whole set of regions. The distributions for size and partial inclusion are similarly estimated, extracting features which are both local and global as a result.

Decision. The decision step selects the most significant rules given the current state of the feature space (see Sec. 2.3), deciding how to combine the extracted dissimilarity features in a unique dissimilarity measure ($D^k(i,j)$) for each pair of connected regions. Finally, the most similar pair of regions is selected for merging.

As presented in Fig. 2, feature extraction and decision are iterated in successive mergings until the whole image support is represented by a single region.

The tracking of successive merging leads to an optimized BPT image representation spanning a set of mergings from the initial partition (leave nodes) to the root of the BPT.

2.3 Combining Multiple Rules

The main difficulty of using higher level features is how to deal with multiple merging rules. For this purpose we estimate the distribution of the dissimilarity values associated to each rule over the whole image. The idea behind this, is to put in context the significance of the local dissimilarity values computed over pairs of regions.

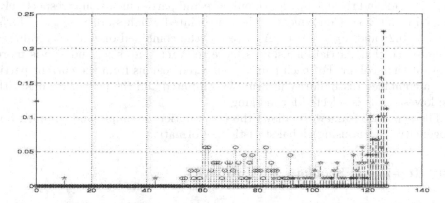

Fig. 3. Pdf estimation of the dissimilarity values associated to each rule for the image shown in Fig. 8. Colour, size and partial inclusion are represented respectively by circles, starts and asterisks.

To estimate the probability density function (pdf) we compute the histogram and divide by the total number of measures. The values of the dissimilarity measures are normalized to the interval $[0, 1]$, and quantified with a sufficient number of bins using a non-linear quantizer. In order to give more relevance to the lower dissimilarity values, we expand this part of the axis and compress the upper part, since lowest values represent rules which tend to be true.

We assume that a **uniform distribution of the dissimilarity values of a rule do not provide relevant information regarding this rule**.

For this purpose we estimate the information of each rule by computing its entropy as, $H_r = -\sum_i p_i log_2(p_i)$. Where p_i is the probability of occurrence of a dissimilarity measure for a given rule. The entropy is the statistical descriptor being computed, and allows combining the individual dissimilarity values in the k-*the* iteration associated to a pair of regions (i, j) in a unique weighted dissimilarity value as

$$D^k(i, j) = \sum_c w_c^k d_c^k(i, j),$$

where $w_c^k = \dfrac{H_c^{max} - H_c}{\sum_{\forall ruler} H_r^{max} - H_r}.$

Note that $\sum_c w_c = 1$ and since $d(i,j) \in [0,1]$ the combined value is also in the interval $[0,1]$. Entropy is maximal for uniform distributions, thus w_c^k tends to zero decreasing the contribution of this rule in such case.

For example, the image shown in the first row of Fig. 8(a) is over-segmented to provide a set of 49 regions, see 8(b). In this case, there are 89 connected pairs of regions in the first iteration. The rules are assessed providing the dissimilarity measures which are used to estimate the pdf for the aforementioned rules (colour, size, and partial-inclusion). For the initial partition, the distribution of the partial-inclusion dissimilarity values presented in Fig. 3, shows that regions are either totally included or not included.

The distribution of the colour and size values are flatter than the partial-inclusion pdf and the entropy for colour, size and partial inclusion is respectively, 5.0, 4.74 and 3.34. Consequently, their associated weights are $w_{Colour} = 0.25$, $w_{Size} = 0.28$ and $w_{Inc} = 0.46$. As a result, the combined dissimilarity value is weighted so that partial-inclusion is more relevant than size, and size is more relevant than colour. For each pair of connected regions from the current partition a combined dissimilarity measure is computed and the pair of regions with the lowest value is selected for merging.

The syntactic framework allows the combination of simple and complex homogeneity criteria using global statistical information.

2.4   Top-Down: Semantic BPT Analysis

In the previous motivation section, it has been stated that multiple detections of a single instance is a common problem in image analysis applications based on BPTs. This paper proposes a technique to cope with these cases and choose among all the BPT nodes which are candidate to contain a semantic instance of a class.

Before formulating the selection criterion, it is necessary to introduce the concept of **BPT Semantic Neighborhood**. A BPT Semantic Neighborhood is a subset of connected BPT nodes that represent instances of the same semantic class. Notice that a BPT Semantic Neighborhood is associated to a specific semantic class and that a BPT node could represent more than an instance of different classes. Figure 4 shows and example of two BPT semantic neighborhoods of the same class "A".

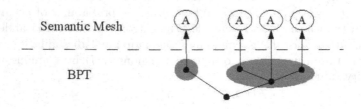

Fig. 4. BPT semantic neighborhood

Simple Classes. We define as **simple** classes those ones whose instances can be completely represented by a single BPT node. They are modelled according to the perceptual caracteristics of the region, for example, as a list of low-level descriptors.

In the simple classes case, the selection criterion is based on the supposition that all instances of the class associated to a BPT Semantic Neighborhood represent in fact the same instance. Among all the candidates, the most similar one to the perceptual model will be chosen and the rest discarded.

Composite Classes. We consider **composite** classes those which are defined as a combination of semantic instances (SI) of other classes that satisfy certain semantic relations (SR). The model of a composite class includes instances of lower level classes, which, at the same time, are described by other simple and/or composite models. As shown in Fig. 5 a), the semantic decomposition can be iterated until reaching the lowest possible level, corresponding to simple classes. Such top-down expansion can be summarized in a basic graph called Semantic Tree (ST), as shown in Fig. 5 b).

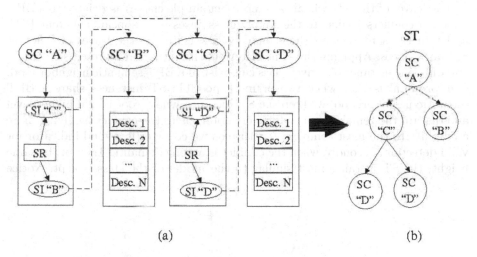

(a)   (b)

Fig. 5. a) Semantic decomposition, b) Semantic Tree (ST)

When working with composite classes, the detection algorithm will look for combinations of the detected instances that satisfy the relations represented by the model. For each valid combination, a new ST node candidate is created and linked to those ST nodes representing the composing instances. The hierarchical decomposition of classes drives to a recursive detection algorithm [6] according to a bottom-up expansion of the ST.

Towards the final goal of building instances of Semantic Trees, the algorithm prevents creating unnecessary links among ST nodes. Every time a new ST node candidate is added to the Semantic Mesh, it must become the root of the

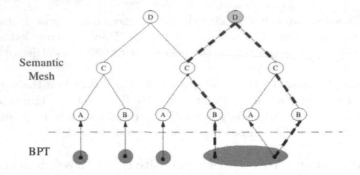

Fig. 6. Cycles discard ST nodes candidates in gray

inferior tree structure. That is, the addition of a new node must not close any cycle through the lower levels of the Semantic Mesh. This condition must also be assessed through the BPT semantic neighborhoods. Figure 6 shows a case in which the ST nodes in gray are discarded due to a cycle through a BPT Semantic Neighborhood of class "B".

The leaves of the STs will always represent simple classes associated to a BPT node, so conflicts similar to the simple class cases may appear. Different BPT nodes in the same neighborhood may be associated to the leaves of different Semantic Trees. Applying the same criterion, this situation represents a conflict and only one instance of a given class can exist in a BPT semantic neighborhood. Our proposal is that when two or more possible ST instances share a BPT semantic neighborhood, we keep the ST instance whose root is in a higher level and discard the remaining ones. This criterion solves the conflict by giving more credibility to the most complex structure as we consider it a good indicator for valid detection. Secondly, when the conflict is among Semantic Trees of the same height, the ST instance most similar to the model of the class is kept. Notice

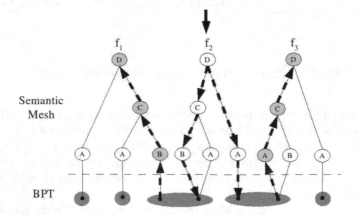

Fig. 7. Consolidated ST ($f2 > f1$ and $f2 > f3$) overlaps with nodes in gray through the BPT semantic neighborhood and forces their deletion

that discarding model is the one represented at the root of the Semantic Tree. Figure 7 shows an example in which the Semantic Tree with probability f_2 is kept and the other two discarded.

3   Results

This section presents two applications of the enhanced BPT framework. An example of traffic sign detection illustrates in detail the framework and the proposed framework demonstrates its suitability for a general application performing the detection of laptops in a smart room.

For both applications the initial partition was automatically created using a colour homogeneity criterion in the YUV space with weights 1, 0.5 and 0.5, respectively and a PSNR of 23 dB for road sign detection and 24dB for laptop detection.

3.1   Application 1: Roadsign Detection

The results presented in this application were generated from images from the CLIC database [7] that contains a traffic sign.

The images shown in Fig. 8 present three examples of object extraction using the enhanced BPT framework. As we can see, column a) is the original image, column b) its partition represented by the mean colour of each region and column

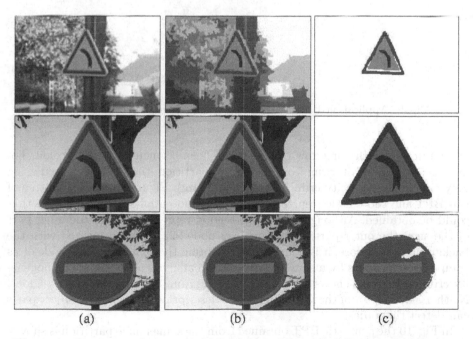

(a)   (b)   (c)

Fig. 8. Application 1: a) Original, b) Initial partition and c) Extracted traffic sign

c) the only detected instance of the object of interest. These examples show the performance of the system for this application.

The detailed analysis of the image shown in the first row is used to exemplify the syntactic and the semantic analysis in the enhanced framework.

Syntactic Analysis. The aim of this section is to show, by means of a visual example, how well the syntactic framework used for the BPT creation can overcome the limitations of simple homogeneity criteria (e.g. colour). For this task, we aim to compare the best object candidate represented as a single node of the BPT that can be obtained using either simple or syntactic homogeneity criteria.

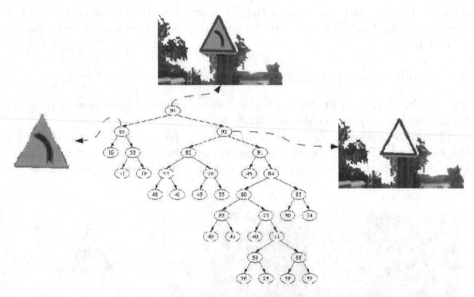

Fig. 9. Application 1: BPT using a simple homogeneity criteria: colour

The image on the first row of Fig. 8 has been segmented into 49 colour homogeneous regions. Starting from this initial partition, the BPT is created using only simple homogeneity criteria (colour and size). Figure 9 shows a subtree of this BPT and those nodes representing the best potential object candidates that could be obtained. Although node 85 is a good representation of the inner part of the sign, the outer part, represented by node 92, has been merged with the background. Therefore, it is not possible to obtain the sign as a single node or as composition of nodes by means of thetop-down analysis using simple homogeneity criteria. Node 93 shows the region resulting from merging the above regions. Neither a descriptor of the whole sign nor a description based on the object parts can detect this object.

In Fig. 10 the syntactic BPT obtained from the same image partition is shown. This BPT is computed using a combination of simple (colour, size) and complex

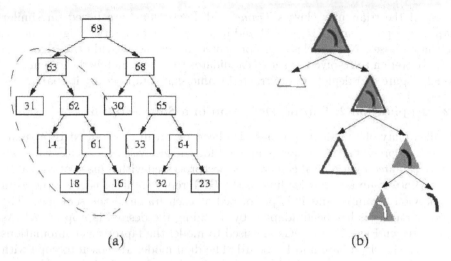

(a)   (b)

Fig. 10. Application 1: a) BPT using multiple criteria: colour, size and partial inclusion, b) Regions associated to the nodes in ellipse.

(syntactic: partial inclusion) criteria. In this case, the sign is represented by a single node which is a better object candidate than the ones found in 9.

Semantic Analysis. The syntactic analysis creates a BPT that, apart from having a node that fully represents the object of interest, it also contains nodes with the parts that composed them. The traffic sign is a type of object that suits its modeling from its parts. The red frame, white background and black silhouette are described separately with colour and shape descriptors, that is, with a perceptual model. Afterwards, references to this classes are used for a semantic modeling of the traffic sign class with a Description Graph [8], as shown in Fig. 11.

<u>Semantic class</u>: Curve Traffic Sign

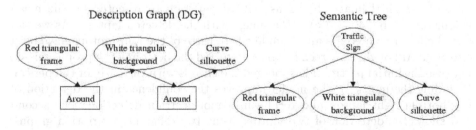

Description Graph (DG)   Semantic Tree

Fig. 11. Application 1: Description Graph and Semantic Tree of the traffic sign

Figure 10 a) shows the syntactic BPT created in the previous section. In Fig. 10 b) there are the regions associated to each BPT node. As the perceptual model of the triangular classes *Frame* and *Background* are based on similar shape descriptors, nodes 14, 61, 62 and 63 are a BPT Semantic Neighborhoods for these classes. Nevertheless, the presented technique provide the algorithm with a criterion to resolve the set of candidates that best matched the semantic model. Figure 8 c) depicts the extracted traffic sign composed by its parts.

3.2 Application 2: Laptop Detection in a Smart Room

The flexibility of the algorithm has also been tested in the context of a smart room. The presented technique is applied for the analysis of a QCIF video sequence acquired by a zenital camera and postprocessed with a mask of the table.

The bottom-up analysis leading to the BPT representation is the same as in the previous example and it is performed at each frame of the sequence. The top-down analysis has been adapted by including the desired laptop model. As we can see on Fig. 12, two DGs are used to model the laptops as combinations of screen, chassis, mouse and keyboard. The dual model id chosen to cope with the important variability introduced by the poor image resolution.

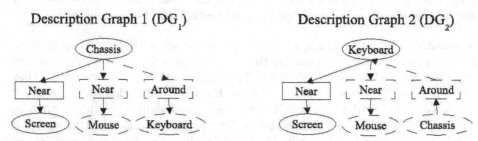

Fig. 12. Application 2: Description Graph and Semantic Tree of the laptop

Fig. 13 shows three examples of laptop detection in the smart room. Column a) is the original image, b) its associated partition represented by the mean colour of each region and c) is the image with the detected laptops. As we can see, the first row presents an example where three of the four laptops have been detected. Although the screen has been detected the bottom left laptop is missed because the initial partition has merged the chassis with the body of the person. Including the mouse in the model improves the confidence in the detection of the bottom right laptop and results in a more certain detection. The second row shows the detection of two laptops using two different description graphs. One of them is based on a chassis and a screen, while the second one is based on the keyboard and the screen. In the second case, the chassis is not detected due to an heterogeneous color on the laptop surface. Notice that screens have been correctly segmented despite the narrow visibility angle and a low contrast compared to the chassis. The last row shows a detection of a laptop with a partial

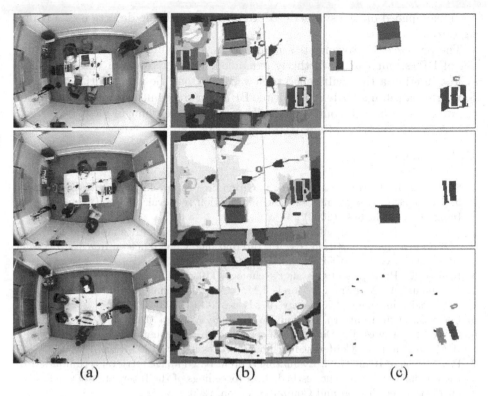

Fig. 13. Applications 2: a) Original image, b) Initial partition and c) Extracted laptop

occlusion of the keyboard. In this case, the segmentation of the screen was pretty bad, creating a region with a distorted shape. Nevertheless, its proximity to the keyboard is enough to consider it a screen in the context of a laptop.

4   Conclusions

The representation of images from the region-based perspective offered by BPTs provides several advantages for analysis applications. However, how to create this BPTs and how to process it is a key issue for a good performance. This paper has presented two enhancements for the classic BPT.

Firstly, the classic merging of BPT nodes based on colour has been enriched by allowing to combine, using statistical information, multiple criteria, such as the syntactic criteria. A practical example has shown a case in which considering colour, size and partial inclusion of regions drives to the creation of better BPTs.

Secondly, the lack of semantic knowledge of the BPTs usually creates more nodes than semantic entities in the content. Some previous assumptions from the semantic point of view have been presented in this paper based on the definition of BPT Semantic Neighbourhoods. The theoretical approach has also been exemplified on the syntactic BPT, which allowed the detection of a traffic sign from its parts.

The proposed improvements are able to enrich the creation and understanding of BPTs, but can be furtherly expanded. Further research will study new syntactic criteria that will generate new BPTs from the same initial partition. New criteria will may drive to multiple BPTs for the same content, which arises the question of how to combine them from a semantic point of view.

References

1. Salembier, P., Garrido, L.: Binary partition tree as an efficient representation for image processing, segmentation, and information retrieval. IEEE Transactions on Image Processing **9**(4) (2000) 561–576
2. Ferran, C., Casas, J.R.: Object representation using colour, shape and structure criteria in a binary partition tree. In: International Conference on Image Processing (ICIP-2005), Genoa, IEEE (2005) III/1144—1151
3. Koffka, K.: Principles of Gestalt Psychology. (1935)
4. Desolneux, A., Moisan, L., Morel, J.M.: Computational gestalts and perception thresholds. Journal of Physiology (97, Issues 2-3,) (2003) 311–322
5. Fu, K.S.: Digital Pattern Recognition. Springer-Verlag (1976)
6. Giró, X., Marqués, F.: Detection of semantic objects using description graphs. In: IEEE International Conference on Image Processing, ICIP'05. (2005)
7. P.A Moellic, P.Hede, G.C.: Evaluating content based image retrieval techniques with the one million images clic testbed. In: Proceedings of the International Conference on Pattern recognition and Computer Vision. (2005)
8. Giró, X., Vilaplana, V., Marqués, F., Salembier, P.: 7. In: Multimedia Content and the Semantic Web. Wiley (2005) 203–221

Semantic Image Analysis Using a Learning Approach and Spatial Context

G.Th. Papadopoulos[1,2], V. Mezaris[2], S. Dasiopoulou[1,2], and I. Kompatsiaris[2]

[1] Information Processing Laboratory
Electrical and Computer Engineering Department
Aristotle University of Thessaloniki, Greece
[2] Informatics and Telematics Institute
1st Km Thermi-Panorama Road
Thessaloniki, GR-57001 Greece
{papad, bmezaris, dasiop, ikom}@iti.gr

Abstract. In this paper, a learning approach coupling Support Vector Machines (SVMs) and a Genetic Algorithm (GA) is presented for knowledge-assisted semantic image analysis in specific domains. Explicitly defined domain knowledge under the proposed approach includes objects of the domain of interest and their spatial relations. SVMs are employed using low-level features to extract implicit information for each object of interest via training in order to provide an initial annotation of the image regions based solely on visual features. To account for the inherent visual information ambiguity spatial context is subsequently exploited. Specifically, fuzzy spatial relations along with the previously computed initial annotations are supplied to a genetic algorithm, which uses them to decide on the globally most plausible annotation. In this work, two different fitness functions for the GA are tested and evaluated. Experiments with outdoor photographs demonstrate the performance of the proposed approaches.

1 Introduction

Recent advances in both hardware and software technologies have resulted in an enormous increase of the images that are available in multimedia databases or over the internet. As a consequence, the need for techniques and tools supporting their effective and efficient manipulation has emerged. To this end, several approaches have been proposed in the literature regarding the tasks of indexing, searching and retrieval of images [1], [2].

The very first attempts to address these issues concentrated on visual similarity assessment via the definition of appropriate quantitative image descriptions, which could be automatically extracted, and suitable metrics in the resulting feature space. Coming one step closer to treating images the way humans do, these were later adapted to a finer granularity level, making use of the output of segmentation techniques applied to the image [1]. Whilst low-level descriptors, metrics and segmentation tools are fundamental building blocks of any image

Y. Avrithis et al. (Eds.): SAMT 2006, LNCS 4306, pp. 199–211, 2006.
© Springer-Verlag Berlin Heidelberg 2006

manipulation technique, they evidently fail to fully capture by themselves the semantics of the visual medium; achieving the latter is a prerequisite for reaching the desired level of efficiency in image manipulation. To this end, research efforts have concentrated on the semantic analysis of images, combining the aforementioned techniques with *a priori* domain specific knowledge, so as to result in a high-level representation of images [2]. Domain specific knowledge is utilized for guiding low-level feature extraction, higher-level descriptor derivation, and symbolic inference.

Depending on the adopted knowledge acquisition and representation process, two types of approaches can be identified in the relevant literature: implicit, realized by machine learning methods, and explicit, realized by model-based approaches. The usage of machine learning techniques has proven to be a robust methodology for discovering complex relationships and interdependencies between numerical image data and the perceptually higher-level concepts. Moreover, these elegantly handle problems of high dimensionality. Among the most commonly adopted machine learning techniques are Neural Networks (NNs), Hidden Markov Models (HMMs), Bayesian Networks (BNs), Support Vector Machines (SVMs) and Genetic Algorithms (GAs) [3], [4]. On the other hand, model-based image analysis approaches make use of prior knowledge in the form of explicitly defined facts, models and rules, i.e. they provide a coherent semantic domain model to support "visual" inference in the specified context [5], [6].

Regardless of the adopted approach to knowledge acquisition, the exploitation of spatial context in the analysis process makes necessary the definition and extraction of context attributes, which are most commonly limited to spatial relations. The relevant literature considers roughly of two categories for the latter: angle-based and projection-based approaches. Angle-based approaches include [14], where a pair of fuzzy k-NN classifiers are trained to differentiate between the *Above/Below* and *Left/Right* relations and the work of [19], where an individual fuzzy membership function is defined for every relation and applied directly to the estimated angle-histogram. Projection-based approaches include [20], where qualitative directional relations in terms of the centre and the sides of the corresponding objects MBRs were defined, and [13], where the use of a representative polygon was introduced.

In this paper, a semantic image analysis approach is proposed that combines two types of learning algorithms, namely SVMs and GAs, with explicitly defined knowledge in the form of an ontology that specifies domain objects and fuzzy spatial relations. SVMs are employed for performing an initial mapping between low-level visual features and the domain objects in the ontology (i.e. generating an initial hypothesis set for every image region) at the region level, whereas a GA is subsequently used to optimize this mapping over the entire image, taking into account the spatial context. Representation of the latter relies on fuzzy spatial relations extraction, building on the principles of both projection- and angle- based methodologies. Application of the proposed approach to images of the specified domain results in the generation of fine granularity semantic representations, i.e. a segmentation map with semantic labels attached to each segment. These initial labels can be used to infer additional knowledge.

The paper is organized as follows: Section 2 presents the overall system architecture. Sections 3 and 4 describe the employed low- and high-level knowledge respectively. Sections 5 and 6 detail individual system components. Experimental results for outdoor photographs are presented in Sect. 7 and conclusion are drawn in Sect. 8.

2   System Overview

The overall architecture of the proposed system for semantic image analysis is illustrated in Fig. 1. Initially, a segmentation algorithm is applied in order to divide the given image into regions, which are likely to represent meaningful semantic objects. Then, for every resulting segment, low-level descriptions and spatial relations are estimated, the latter according to the relations supported by the domain ontology.

Estimated low-level descriptions for each region are employed for generating initial hypotheses regarding the region's semantic label. This is realized by evaluating the compound low-level descriptor vector by a set of SVMs, each trained to identify instances of a single concept defined in the ontology. SVMs were selected for this task due to their generalization ability and their efficiency in solving high-dimensionality pattern recognition problems [7], [8].

The generated hypothesis sets for each region with the associated degrees of confidence for each hypothesis along with the spatial relations computed for every image segment, are subsequently employed for selecting a globally optimal set of semantic labels for the image regions by introducing them to a genetic algorithm. The choice of a GA for this task is based on its extensive use in a wide variety of global optimization problems [9], where they have been shown to outperform traditional methods, and is further endorsed by the authors previous experience [10], which showed promising results.

3   Low-Level Visual Information Processing

3.1   Segmentation and Feature Extraction

In order to implement the initial hypothesis generation procedure, the examined image has to be segmented into regions and suitable low-level descriptions have to be extracted for every resulting segment. In the current implementation, an extension of the Recursive Shortest Spanning Tree (RSST) algorithm has been used for segmenting the image [11].

Considering low-level descriptions, specific descriptors of the MPEG-7 standard have been selected, namely the *Scalable Color*, *Homogeneous Texture*, *Region Shape* and *Edge Histogram* descriptors. Their extraction is performed according to the guidelines provided by the MPEG-7 eXperimentation Model (XM) [12]. The above descriptors are extracted for every computed image segment and are

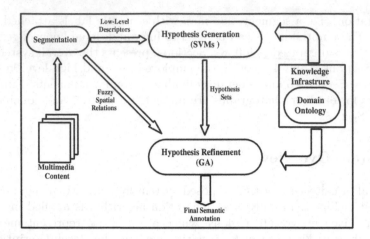

Fig. 1. System Architecture

combined in a single feature vector. This vector constitutes the input to the SVMs framework which computes the initial hypothesis set for every segment, as will be described in Sect. 5.

3.2 Spatial Context

Exploiting spatial context, i.e. domain-specific spatial knowledge, in image analysis tasks is a common practice among the object recognition community. It is generally observed that objects tend to be present in a scene within a particular spatial context and thus spatial information can substantially assist in discriminating between objects exhibiting similar visual characteristics. Among the most commonly adopted spatial context representations, directional spatial relations have received particular interest. They are used to denote the order of objects in space. In the present analysis framework, eight fuzzy directional relations are supported, namely *Above* (A), *Right* (R), *Below* (B), *Left* (L), *Below-Right* (BR), *Below-Left* (BL), *Above-Right* (AR) and *Above-Left* (AL).

Fuzzy directional relations extraction in the proposed analysis approach builds on the principles of projection- and angle- based methodologies [13], [14] and consists of the following steps. First, a *reduced box* is computed from the *ground* object's (the object used as reference and is pointed in dark grey in Fig. 2) Minimum Bounding Rectangle (MBR) so as to include the object in a more representative way. The computation of this *reduced box* is performed in terms of the MBR compactness value c, which is defined as the value of the fraction of the objects's area to the area of the respective MBR: If the initially computed c is below a threshold T, the ground objects's MBR is reduced repeatedly until the desired threshold is satisfied. Then, eight cone-shaped regions are formed on top of this reduced box, as illustrated in Fig. 2, each corresponding to one of the defined directional relations. The percentage of the *figure* object (whose relative position is to be estimated and is pointed in light grey in Fig. 2) points that are included in each of the cone-shaped regions determines the degree to which the corresponding directional relation is satisfied. After extensive experimentation, the value of threshold T was set equal to 0.85.

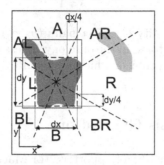

Fig. 2. Fuzzy directional relations definition

4 Knowledge Infrastructure

Among the possible domain knowledge representations, ontologies [15] present a number of advantages, the most important being that they provide a formal framework for supporting explicit, machine-processable semantics definition and they enable the derivation of new knowledge through automated inference. Thus, ontologies are suitable for expressing multimedia content semantics so that automatic semantic analysis and further processing of the extracted semantic descriptions is allowed. Following these considerations, a domain ontology was developed for representing the knowledge components that need to be explicitly defined under the proposed approach. This contains the semantic concepts that are of interest in the examined domain (e.g. in the beach vacation domain: Sea, Sand, Person, etc.), as well as their spatial relations. The values of the latter for the concepts of the given domain, as opposed to concepts themselves that are manually defined, are estimated according to the following ontology population procedure:

Let $S = \{s_i \ , \ i = 1, ..., \mathcal{I}\}$ denote the set of regions produced for an image by segmentation, $O = \{o_p \ , \ p = 1, ... \ \mathcal{P}\}$ denote the set of objects defined in the employed domain ontology and

$$R = \{r_k, \ k = 1, ..., K\} = \{ \ A, \ AL, \ AR, \ B, \ BL, \ BR, \ L, \ R \ \} \ , \qquad (1)$$

denote the set of supported spatial relations. Then, the degree to which s_i satisfies relation r_k with respect to s_j can be denoted as $I_{r_k}(s_i, s_j)$, where the values of function I_{r_k} are estimated according to the procedure of Sect. 3.2 and belong to $[0, 1]$. To populate the ontology, this function needs to be evaluated over a set of segmented images with ground truth annotations, that serves as a training set. More specifically, the mean values, $I_{r_k mean}$, of I_{r_k} are estimated, for every k

over all region pairs of segments assigned to objects (o_p, o_q), $p \neq q$. Additionally, the variance values $\sigma^2_{r_k}$ (2), are obtained for each of the relations.

$$\sigma^2_{r_k} = \frac{\sum_{i=1}^{\mathcal{N}}(I_{r_k i} - I_{r_k mean})^2}{\mathcal{N}} \quad , \tag{2}$$

where $\mathcal{N}$ denotes the plurality of object pairs (o_p, o_q) for which r_k is satisfied. The calculated values are stored in the ontology. These constitute the constraints input to the optimization problem which is solved by the genetic algorithm, as will be described in Sect. 6.

5   Initial Hypothesis Generation

As already described in Sect. 2, a Support Vector Machines (SVMs) structure is utilized to compute the initial hypothesis set for every image segment. Specifically, an individual SVM is introduced for every defined concept of the employed domain ontology, to detect the corresponding instances. Each SVM is trained under the 'one-against-all' approach. For that purpose, the training set assembled in Sect. 4 is employed and the combined region feature vector, as defined in Sect. 3.1, constitutes the input to each SVM. For the purpose of initial hypothesis generation, every SVM returns a numerical value in the range $[0, 1]$ which denotes the degree of confidence to which the corresponding segment is assigned to the concept associated with the particular SVM. The metric adopted is defined as follows: For every input feature vector the distance D from the corresponding SVM's separating hyperplane is initially calculated. This distance is positive in case of correct classification and negative otherwise. Then, a sigmoid function [16] is employed to compute the degree of confidence, DOC, as follows:

$$DOC = \frac{1}{1 + e^{-mD}} \quad , \tag{3}$$

where the slope parameter m is experimentally set. The pairs of all domain concepts and their respective degree of confidence comprise each segment's hypothesis set. The above SVM structure was realized using the SVM software libraries of [17].

6   Hypothesis Refinement

As outlined in Sect. 2, after the initial set of hypotheses is generated (Sect. 5), based solely on visual features, and the fuzzy spatial relations are computed for every pair of image segments (Sect. 3.2), a genetic algorithm (GA) is introduced to decide on the optimal image interpretation. The GA is employed to solve a global optimization problem, while exploiting the available domain spatial knowledge, and thus overcoming the inherent visual information ambiguity. Spatial knowledge is obtained as described in Sect. 4 and the resulting learnt fuzzy spatial relations serve as constraints denoting the "allowed" domain objects spatial topology.

Fitness function

More specifically, the proposed algorithm uses as input the initial hypothesis sets (generated by the SVMs structure), the fuzzy spatial relations extracted between the examined image segments, and the spatial-related domain knowledge as produced by the particular training process. Under the proposed approach, each chromosome represents a possible solution. Consequently, the number of the genes comprising each chromosome equals the number $\mathcal{I}$ of the segments s_i produced by the segmentation algorithm and each gene assigns a defined domain concept to an image segment.

An appropriate *fitness function* is introduced to provide a quantitative measure of each solution fitness, i.e. to determine the degree to which each interpretation is plausible:

$$f(C) \;=\; \lambda \times FS_{norm} \;+\; (1 - \lambda) \times SC_{norm} \;, \tag{4}$$

where C denotes a particular chromosome, FS_{norm} refers to the degree of low-level descriptors matching, and SC_{norm} stands for the degree of consistency with respect to the provided spatial domain knowledge. The variable λ is introduced to adjust the degree to which visual features matching and spatial relations consistency should affect the final outcome.

The value of FS_{norm} is computed as follows:

$$FS_{norm} \;=\; \frac{\sum_{i=1}^{\mathcal{I}} I_M(g_{ij}) - I_{min}}{I_{max} - I_{min}} \;, \tag{5}$$

where

$$I_M(g_{ij}) \;\equiv\; DOC_{ij} \;, \tag{6}$$

denotes the degree to which the visual descriptors extracted for segment s_i match the ones of object o_p, where g_{ip} represents the particular assignment of o_p to s_i. Thus, $I_M(g_{ip})$ gives the degree of confidence, DOC_{ip}, associated with each hypothesis and takes values in the interval $[0, 1]$. $I_{min} = \sum_{i=1}^{\mathcal{I}} \min_j I_M(g_{ij})$ is the sum of the minimum degrees of confidence assigned to each region hypotheses set and $I_{max} = \sum_{i=1}^{\mathcal{I}} \max_j I_M(g_{ij})$ is the sum of the maximum degrees of confidence values respectively. For the computation of SC_{norm}, two different approaches are followed, as described in the following subsections.

First approach to fuzzy spatial constraints verification

Estimating the degree to which the spatial constraints between two object to segment mappings g_{ip}, g_{jq} are satisfied is a prerequisite for exploiting spatial context in the analysis procedure. In this work, this degree of satisfaction is expressed by the function $I_s(g_{ip}, g_{jq})$. In the first approach presented in this subsection, I_S (g_{ip}, g_{jq}) is defined with the help of a normalized euclidean distance $d(g_{ip}, g_{jq})$, which is calculated according to the following equation:

$$d(g_{ip}, g_{jq}) = \frac{\sqrt{\sum_{k=1}^{8}(I_{r_k mean}(o_p, o_q) - I_{r_k}(s_i, s_j))^2}}{\sqrt{8}} \quad , \qquad (7)$$

where $I_{r_k mean}$ is part of the knowledge infrastructure, as discussed in Sect. 4, $I_{r_k}(s_i, s_j)$ denotes the degree to which spatial relation r_k is verified for a certain pair of segments s_i, s_j of the examined image and o_p, o_q denote the domain defined concepts assigned to them respectively. Distance $d(g_{ip}, g_{jq})$ receives values in the interval $[0, 1]$. The function I_S (g_{ip}, g_{jq}) is then defined as:

$$I_S\ (g_{ip},\ g_{jq})\ =\ 1 - d(g_{ip}, g_{jq}) \qquad (8)$$

and takes values in the interval $[0, 1]$ as well, where '1' denotes an allowable relation and '0' denotes an unacceptable one. Using this, the values of SC_{norm} is computed according to the following equation:

$$SC_{norm}\ =\ \frac{\sum_{l=1}^{W} I_{S_l}(g_{ij}, g_{pq})}{W} \quad , \qquad (9)$$

where W denotes the number of the constraints that had to be examined.

Second approach to fuzzy spatial constraints verification factor

In this second approach, function $I_S(g_{ip},\ g_{jq})$ returning the degree to which the spatial constraint between the objects involved in the g_{ip} and g_{jq} mappings is satisfied, is set to receive values in the interval $[-1, 1]$, where '1' denotes an allowable relation and '-1' denotes an unacceptable one based on the learnt spatial constraints. To calculate this value for a specific pair of objects o_i and o_j the following procedure is used. For every computed triplet $[r_k\ I_{r_k mean}\ \sigma_{r_k}^2]$ of the corresponding spatial constraint (Sect. 4) where $I_{r_k mean} \neq 0$, a triangular fuzzy membership function [21] is formed, as illustrated in Fig. 3(a), to compute the corresponding degree.

Let d_k denote the value resulted from applying the membership function with respect to relation r_k, as illustrated in Fig. 3(b). Once, the corresponding d_k values have been computed for each relation, i.e. for $k = 0, 1, ..8$, they are combined to form the degree $I_S(g_{ip},\ g_{jq})$ to which the corresponding spatial constraint is satisfied. The calculation is based on based on (10), where r_m are the relations for which $I_{r_m mean} \neq 0$, and r_n are the extracted relations for which $I_{r_n mean} = 0$.

$$I_S\ (g_{ip},\ g_{jq}) = \frac{\sum_m \frac{I_{r_m} * d_m}{IR}}{\sum_m d_m} - \sum_n I_{r_n}\ , \quad i \neq j \qquad (10)$$

where IR stands for

$$IR = \sum_m I_{r_m} \qquad (11)$$

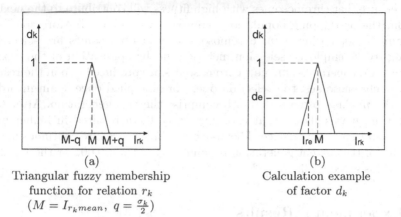

(a)
Triangular fuzzy membership
function for relation r_k
($M = I_{r_k mean}$, $q = \frac{\sigma_k}{2}$)

(b)
Calculation example
of factor d_k

Fig. 3. Triangular fuzzy membership function

Then, SC_{norm} is calculated according to the following equation:

$$SC_{norm} = \frac{SC + 1}{2} \quad and \quad SC = \frac{\sum_{l=1}^{W} I_{S_l}(g_{ij}, g_{pq})}{W}, \tag{12}$$

where W denotes again the number of the constraints that had to be examined.

Implementation of the genetic algorithm

To implement the previously described optimization process, a population of 200 chromosomes is employed, and it is initialized with respect to the input set of hypotheses. After the population initialization, new generations are iteratively produced until the optimal solution is reached. Each generation results from the current one through the application of the following operators.

- Selection: a pair of chromosomes from the current generation are selected to serve as parents for the next generation. In the proposed framework, the Tournament Selection Operator [18], with replacement, is used.
- Crossover: two selected chromosomes serve as parents for the computation of two new offsprings. Uniform crossover with probability of 0.7 is used.
- Mutation: every gene of the processed offspring chromosome is likely to be mutated with probability of 0.008. If mutation occurs for a particular gene, then its corresponding value is modified, while keeping unchanged the degree of confidence.

Parameter λ, regulating the relative weights of low-level descriptor matching and spatial context consistency was set to 0.35 after experimentation. The resulting weight of SC_{norm}, points out the importance of spatial context in the optimization process.

To ensure that chromosomes with high fitness will contribute to the next generation, the overlapping populations approach was adopted. More specifically, assuming a population of m chromosomes, m_s chromosomes are selected according to the employed selection method, and by application of the crossover and mutation operators, m_s new chromosomes are produced. Upon the resulting $m + m_s$ chromosomes, the selection operator is applied once again in order to select the m chromosomes that will comprise the new generation. After experimentation, it was shown that choosing $m_s = 0.4m$ resulted in higher performance and faster convergence. The above iterative procedure continues until the diversity of the current generation is equal to/less than 0.001 or the number of generations exceeds 50.

7   Experimental Results

In this section, we present experimental results from testing the proposed approach in outdoor images. First, a domain ontology had to be developed to represent the domain objects of interest and their relations. Six concepts are currently supported, namely *Sky, Water, Ground, Building, Vegetation* and *Rock*.

Then, a set of 200 randomly selected images belonging to the beach vacation domain were used to assemble a training set for the low-level implicit knowledge acquisition (SVMs training) and computation of the fuzzy spatial constraints. A corresponding set comprising 400 images was similarly formed to serve as a test set for the evaluation of the proposed system performance. Each image of the training/test set was first manually annotated according to the domain ontology definitions.

According to the SVMs training process, a set of instances were selected for every defined domain concept from the assembled training image set. The Gaussian radial basis function was used as a kernel function by each SVM, to allow for nonlinear discrimination of the samples. The low-level combined feature vector, as described in detail in Sect. 3.1, is composed of 433 values, normalized in the interval $[-1, 1]$. On the other hand, for the acquisition of the fuzzy spatial constraints, the procedure described in Sect. 4 was followed for each possible combination of the defined domain objects that were present in the employed image training set.

Based on the trained SVMs structure, initial hypotheses are generated for each image segment as described in Sect. 5, which are then passed in the genetic algorithm along with the fuzzy spatial constraints in order to determine the globally optimal interpretation. In Fig. 4 indicative results are given showing the input image, the annotation resulting from the initial hypotheses set, considering for each image segment the hypothesis with the highest degree of confidence, and the final interpretation after the application of the genetic algorithm.

In Table 1, quantitative performance measures are given in terms of precision and recall for both approaches to fuzzy spatial constraints verification. Table 1 makes evident the superiority of the first out of the two approaches to fitness evaluation. It must be noted that for the numerical evaluation, any object present

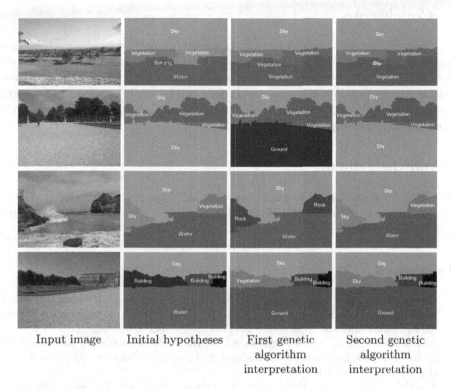

Input image   Initial hypotheses   First genetic   Second genetic
                                         algorithm   algorithm
                                         interpretation   interpretation

Fig. 4. Indicative results for the outdoor images domain

Table 1. Numerical evaluation for the outdoor images domain

	Initial Hypothesis		Final interpretation 1		Final interpretation 2	
Object	precision	recall	precision	recall	precision	recall
Sky	93.33%	76.36%	93.33%	83.17%	93.33%	76.36%
Water	58.54%	42.86%	63.42%	52%	56.10%	46.0%
Building	81.72%	94.95%	87.83%	93.52%	81.74%	94.95%
Rock	61.11%	40.74%	83.33%	57.69%	72.22%	56.52%
Ground	56.52%	61.90%	78.26%	43.90%	69.57%	44.44%
Vegetation	42.47%	65.96%	39.73%	85.29%	41.10%	71.43%
Total precision	71.39 %		75.83 %		72.22 %	

in the examined image test set that was not included in the domain ontology concept definitions, e.g. umbrella, was not taken into account.

After a careful observation of the presented results, we can confirm the good generalization ability of SVMs, regardless of the usage of a limited training set. Furthermore, we can justify the choice of using a genetic algorithm to reach an optimal image interpretation given degrees of confidence for visual similarity and spatial consistency against the domain definitions.

8   Conclusions

In this paper, an approach to knowledge-assisted semantic image analysis that couples Support Vector Machines (SVMs) with a Genetic Algorithm (GA) is presented. The proposed system was tested with outdoor photographs and produced satisfactory results. Furthermore, the system can be easily applied to additional domains, given the fact that an appropriate domain ontology is defined and the corresponding training sets are formed.

Acknowledgements

The work presented in this paper was partially supported by the European Commission under contracts FP6-001765 aceMedia, FP6-027026 K-Space and COST 292 action.

References

1. Smeulders, A. W. M., Worring, M., Santini, S., Gupta, A., Jain, R.: Content-based image retrieval at the end of the early years. IEEE Transactions on Pattern Analysis and Machine Intelligence 22(12) (2000) 1349–1380
2. Al-Khatib, W., Day, Y. F., Ghafoor, A., Berra, P. B.: Semantic Annotation of Images and Videos for Multimedia Analysis. 2nd European Semantic Web Conference (ESWC) Greece (2005)
3. Assfalg, J., Berlini, M., Del Bimbo, A., Nunziat, W., Pala, P.: Soccer Highlights Detection and Recognition using HMMs. IEEE International Conference on Multimedia & Expo (ICME) (2005) 825–828
4. Zhang, L., Lin, F. Z., Zhang, B.: Support Vector Machine Learning for Image Retrieval. International Conference on Image Processing (2001)
5. Dasiopoulou, S., Mezaris, V., Papastathis, V. K., Kompatsiaris, I., Strintzis, M.G.: Knowledge-Assisted Semantic Video Object Detection. IEEE Transactions on Circuits and Systems for Video Technology, Special Issue on Analysis and Understanding for Video Adaptation, 15(10) (2005) 1210-1224
6. Hollink, L., Little, S., Hunter, J.: Evaluating the Application of Semantic Inferencing Rules to Image Annotation. 3rd International Conference on Knowledge Capture (K-CAP05) Canada (2005)
7. Kim, K. I., Jung, K., Park, S. H., Kim, H. J.: Support vector machines for texture classification. IEEE Trans. on Pattern Analysis and Machine Intelligence 24(11) (2002) 1542–1550
8. Chapelle, O., Haffner, P., Vapnik, V.: Support vector machines for histogram-based image classification. IEEE Transactions on Neural Networks 10(5) (1999) 1055-1064
9. Mitchell, M.: An introduction to genetic algorithms. MIT Press (1995)
10. Voisine, N., Dasiopoulou, S., Precioso, F., Mezaris, V., Kompatsiaris, I., Strintzis, M. G.: A Genetic Algorithm-based Approach to Knowledge-assisted Video Analysis. Proc. IEEE International Conference on Image Processing (ICIP) Genova (2005)
11. Adamek, T., O'Connor, N., Murphy, N.: Region-based Segmentation of Images Using Syntactic Visual Features Workshop on Image Analysis for Multimedia Interactive Services (WIAMIS) Montreux Switzerland (2005)

12. MPEG-7 Visual Experimentation Model (XM). Version 10.0 ISO/IEC/JTC1/ SC29/WG11 Doc. N4062 (2001)
13. Skiadopoulos, S., Giannoukos, C., Sarkas, N., Vassiliadis, P., Sellis,T., Koubarakis, M.: 2D topological and direction relations in the world of minimum bounding circles. IEEE Transactions on Knowledge and Data Engineering 17(12) (2005) 1610–1623
14. Wang, Y., Makedon, F., Ford, J., Shen, L., Golding, D.: Generating Fuzzy Semantic Metadata Describing Spatial Relations from Images using the R-Histogram. JCDL '04 June 7-11 Arizona USA (2004)
15. Staab, S., Studer, R.: Handbook on ontologies. in Int. Handbooks on Information Systems Berlin Germany Springer-Verlag (2004)
16. Tax, D., Duin, R.: Using two-class classifiers for multi-class classification. in Proc. Int. Conf. Pattern Recognition Quebec City Canada vol. 2 (2002) 124-127
17. Chang, C.-C., Lin, C.-J.: LIBSVM: A library for support vector machines. http://www.csie.ntu.edu.tw/ cjlin/libsvm (2001)
18. Goldberg, D., Deb, K.: A comparative analysis of selection schemes used in genetic algorithms. In Foundations of Genetic Algorithms G. Rawlins (1991) 69–93
19. Millet, C., Bloch, I., Hede, P., Moellic, P.-A.: Using relative spatial relationships to improve individual region recognition. EWIMT London (2005)
20. Hollink, L., Nguyen, G., Schreiber, G., Wielemaker, J., Wielinga, B., Worring, M.: Adding Spatial Semantics to Image Annotations. Proceedings of International Workshop on Knowledge Markup and Semantic Annotation (2004)
21. Cox, E.: Fuzzy fundamentals. Spectrum IEEE 29(10) (1992) 58 61

A Context-Based Region Labeling Approach for Semantic Image Segmentation*

Thanos Athanasiadis, Phivos Mylonas, and Yannis Avrithis

School of Electrical and Computer Engineering
National Technical University of Athens
9, Iroon Polytechniou Str.,
157 73 Zographou, Athens, Greece
{thanos, fmylonas, iavr}@image.ntua.gr

Abstract. In this paper we present a framework for simultaneous image segmentation and region labeling leading to automatic image annotation. The proposed framework operates at semantic level using possible semantic labels to make decisions on handling image regions instead of visual features used traditionally. In order to stress its independence of a specific image segmentation approach we applied our idea on two region growing algorithms, i.e. watershed and recursive shortest spanning tree. Additionally we exploit the notion of visual context by employing fuzzy algebra and ontological taxonomic knowledge representation, incorporating in this way global information and improving region interpretation. In this process, semantic region growing labeling results are being re-adjusted appropriately, utilizing contextual knowledge in the form of domain-specific semantic concepts and relations. The performance of the overall methodology is demonstrated on a real-life still image dataset from the popular domains of beach holidays and motorsports.

1 Introduction

Automatic segmentation of images is a very challenging task in computer vision and one of the most crucial steps toward image understanding. A variety of applications such as object recognition, image annotation, image coding and image indexing, utilize at some point a segmentation algorithm and their performance depends highly on the quality of the latter. Comparatively to the research efforts in automatic image and video segmentation [8], [18] and global [9], [14] or region-based [3], [13] image classification, still, human vision perception outperforms state-of-the-art computer algorithms. The main reason for this is that human vision is additionally based on high level a priori knowledge about the semantic meaning of the objects that compose the image and on contextual knowledge about their relationships. Moreover, erroneous image segmentation leads to poor results in recognition of materials and

---

* This research was partially supported by the European Commission under contract FP6-001765 aceMedia and contract FP6-027026 K-SPACE and by the Greek Secretariat of Research and Technology (PENED Ontomedia 03 EΔ 475).

Y. Avrithis et al. (Eds.): SAMT 2006, LNCS 4306, pp. 212–225, 2006.
© Springer-Verlag Berlin Heidelberg 2006

objects, while at the same time, imperfections of global image classification are responsible for deficient segmentation. It is rather obvious that limitations of one prohibit the efficient operation of the other.

In this work we propose an algorithm that involves simultaneously segmentation and detection of simple objects, imitating partly the way that human vision works. An initial region labeling is performed based on matching region's low-level descriptors with concepts stored in an ontological knowledge base; in this way, each region is associated to a fuzzy set of candidate concepts. A merging process is performed based on new similarity measures and merging criteria that are defined at the semantic level with the use of fuzzy sets operations. Our approach can be applied to every region growing segmentation algorithm, like morphological watershed [7], RSST [16], color-edge based and seeded region growing [11], etc., given some necessary modifications. Region growing algorithms start from an initial partition of the image and then an iteration of region merging begins, based on similarity measures until the predefined termination criteria are met. We adjust appropriately these merging process as well as the termination criteria.

We also propose a context representation approach to use on top of semantic region growing. We introduce a methodology to improve the results of image segmentation, based on contextual information. A novel ontological representation for context is utilized, combining fuzzy theory and fuzzy algebra [12] with characteristics derived from the Semantic Web, like the statement's reification technique [21]. In this process, membership degrees of concepts assigned to regions derived by the semantic segmentation process are optimized, according to a context-based membership degree readjustment algorithm. This algorithm utilizes ontological knowledge, in order to provide optimized membership degrees of detected concepts of each region in the scene. Our research efforts employ contextual knowledge derived from the popular domains of beach holidays and motorsports.

The outline of the paper is as follows: Section 2 is dedicated to the knowledge representation used, including the necessary notation used throughout the paper. Section 3 describes the semantic region growing approach of segmentation, examining in detail two variations. Utilization of contextual knowledge is discussed in section 4 and finally section 5 presents the dataset and methodology of the experiments and the results of the proposed algorithms.

2 Knowledge Representation

2.1 Ontology Fuzzification and Fuzzy Relations

The first thing to consider within the proposed approach of semantic image segmentation and labeling is what type of knowledge model to use to describe the contextual information. The latter plays a key role in optimizing the results of both methodologies and is built on a novel ontological representation for context. In general, one possible way to describe ontologies [10] can be formalized as:

$$O = \{C, \{R_{c_i, c_j}\}\} \tag{1}$$

O is an ontology, C is the set of concepts described by the ontology, c_i and c_j are two concepts $c_i, c_j \in C$ and $R_{c_i,c_j} : C \times C \rightarrow \{0,1\}$ is the semantic relation amongst these concepts, as the latter is defined within the semantic framework of the MPEG-7 description [20]. According to this description narrative worlds depicted by or related to multimedia content are represented by describing semantic concepts together with their relations and attributes [6]. Herein, the proposed knowledge model is based on a set of concepts and relations between them, which form the basic elements towards semantic interpretation. Although almost any type of relation may be included to construct the knowledge representation, the two main categories used are *taxonomic* (i.e. ordering) and *compatibility* (i.e. symmetric) relations. However, compatibility relations fail to assist in the determination of the context and therefore the use of ordering relations is more appropriate for such tasks [1]. Thus, a main challenge is the meaningful utilization of information contained in taxonomic relations for the task of context exploitation within semantic image segmentation and object labeling.

In addition, for a knowledge model to be highly descriptive, it must contain a large number of distinct and diverse relations among concepts. However, in this case available information will be scattered among them, making each one of them inadequate to describe a context in a meaningful way. Thus, the utilized relations need to be combined to provide a view of the knowledge that suffices for context definition and estimation. In this work we utilize three types of relations, whose semantics are defined in MPEG-7 [19], namely the *specialization* relation Sp, the *part* relation P and the *property* relation Pr. When modeling real-life information that is governed by uncertainty and fuzziness, fuzzy relations have been proposed to handle such issues. In particular, the above commonly encountered relations can be modeled as fuzzy ordering relations and can be combined for the generation of a meaningful fuzzy, taxonomic relation. Consequently, to tackle such types of relations we propose a "fuzzification" of the previous ontology definition, as follows:

$$O_F = \{C, \{r_{c_i,c_j}\}\}, \text{ where } r_{c_i,c_j} = F(R_{c_i,c_j}) : C \times C \rightarrow [0,1] \qquad (2)$$

In equation (2), O_F defines a "fuzzified" ontology, C is again the set of all possible concepts it describes and r_{c_i,c_j} denotes a fuzzy relation amongst the two concepts $c_i, c_j \in C$. More specifically, given a universe U a crisp set C is described by a membership function $\mu_C : U \rightarrow \{0,1\}$, whereas according to [12], a *fuzzy set F* on C is described by a membership function $\mu_F : C \rightarrow [0,1]$. We may describe the fuzzy set F using the sum notation:

$$F = \sum_{i=1}^{n} c_i / w_i = \{c_1 / w_1, c_2 / w_2, ..., c_n / w_n\} \qquad (3)$$

Where $n = |C|$ is the cardinality of C and $w_i = \mu_F(c_i)$. As in [12], a *fuzzy relation* on C is a function $r_{c_i,c_j} : C \times C \rightarrow [0,1]$ and its *inverse* relation is defined as $r_{c_i,c_j}^{-1} = r_{c_j,c_i}$. Based on the relations r_{c_i,c_j} and, for the purpose of image analysis, we construct the following relation T with use of the above set of fuzzy taxonomic relations: Sp, P and Pr.

$$T = Tr'(Sp \cup P^{-1} \cup Pr^{-1}) \qquad (4)$$

Transitive closure Tr' is required in order for T to be taxonomic, as the union of transitive relations is not necessarily transitive [2].

2.2 Graph Representation of an Image

An image can be described as a structured set of individual objects, allowing thus a straightforward mapping to a graph structure. In this fashion, many image analysis problems can be considered as graph theory problems, inheriting the solid theoretical grounds of the latter. Attributed Relation Graph (ARG) is a type of graph often used in computer vision and image analysis for the representation of structured objects.

Formally, an ARG is defined by spatial entities represented as a set of vertices V and binary spatial relationships represented as a set of edges E: $ARG \equiv \langle V, E \rangle$. Letting G be the set of all connected, non-overlapping regions/segments of an image, then a region $a \in G$ of the image is represented in the graph by vertex $v_a \in V$, where $v_a \equiv \langle a, D_a, L_a \rangle$. More specifically, $D_a = [DC_a \ HT_a]$ is the ordered set of two MPEG-7 Visual Descriptors characterizing the region in terms of low level features, namely *Dominant Color* (*DC*) and *Homogeneous Texture* (*HT*) [15]. Additionally, $L_a = \sum_{i=1}^{|C|} c_i / \mu_a(c_i)$ is the fuzzy set (defined on the crisp set of concepts C, since $c_i \in C$) of candidate concepts for the region, which incorporates the uncertainty of the of the region labeling process.

The adjacency relation between two neighbor regions $a, b \in G$ of the image is represented by graph's edge $e_{ab} \equiv \langle (v_a, v_b), s_{ab} \rangle \in E$. s_{ab} is a similarity value for the pair of adjacent regions (v_a, v_b). This value is calculated based on the semantic similarity of the two regions as described by the two fuzzy sets L_a and L_b:

$$s_{ab} = \max_{c_k \in C}(min(\mu_a(c_k), \mu_b(c_k))), \ a, b \in G \qquad (5)$$

The above formula states that the similarity of two regions is the default fuzzy union (*max*) over all common concepts of the default fuzzy intersection (*min*) of the degrees of membership $\mu_a(c_k)$ and $\mu_b(c_k)$ for the specific concept of the two regions a and b.

Finally, we consider two regions $a, b \in G$ to be connected when at least one pixel of one region is 4-connected to one pixel of the other. In ARG, a neighborhood N_a of a vertex $v_a \in V$ is the set of vertices whose corresponding regions are connected to a: $N_a = \{v_b : e_{ab} \neq \varnothing\}$, $a, b \in G$. It is rather obvious now that the subset of ARG's edges that are incident to region a can be defined as: $E_a = \{e_{ab} : b \in N_a\} \subseteq E$.

The current approach (i.e. using two different graphs within this work) may look unusual to the reader at the first glance; however using RDF to represent our knowledge model does not entail the use of RDF-based graphs for the representation of an image in the image analysis domain. Use of *ARG* is clearly favored for image representation and analysis purposes, whereas RDF-based knowledge model is ideal to store in and retrieve from a knowledge base. The common element of the two representations, which is the one that unifies and strengthens the current approach, is the utilization of a common fuzzy set notation, that bonds together both knowledge models. In the following section we shall focus on the use of the *ARG* model and provide the guidelines for the fundamental initial region labeling of an image.

3 Semantic Region Growing Approach

3.1 Overview

The major target of this work is to improve both image segmentation and recognition of simple objects at the same time, with obvious benefits for problems in the area of image understanding. As mentioned in the introduction, the novelty of the proposed idea lies on blending well established segmentation techniques with mid-level features, in the formal style defined earlier in section 2.2. Our intention is to operate on a higher level of information where regions are linked to concepts rather than only to their visual features. For this purpose a knowledge assisted analysis (KAA) algorithm, discussed in depth in a previous work [4], has been designed and implemented. Population of the fuzzy set L_a for all regions of G, is based on a matching process between the visual descriptors stored in each vertex v_a of the *ARG* and the corresponding visual descriptors of concepts, stored in the form of prototype instances in the corresponding ontological knowledge base.

In order to emphasize that this approach is independent of the selection of the segmentation algorithm, we examine two traditional segmentation techniques, belonging in the general category of region growing algorithms. The first is the watershed segmentation [7], while the second is the Recursive Shortest Spanning tree, also known as RSST [16]. We modify these techniques to operate on the fuzzy sets stored in the *ARG* in a similar way as if they worked on low-level features (such as color, texture, *etc.*) [5]. Both variations follow in principles the algorithmic definition of their traditional counterparts, though several adjustments were considered necessary and were added. We call this overall approach Semantic Region Growing (SRG).

3.2 Semantic Watershed

The watershed algorithm [7] owes its name from the way in which regions are segmented into catchment basins. A catchment basin is the set of points that is the local minimum of a height function (most often the gradient magnitude of the image). After locating these minima, the surrounding regions are incrementally flooded and the places where flood regions touch are the boundaries of the regions. Unfortunately, this strategy leads to oversegmentation of the image; therefore a marker controlled segmentation approach is usually applied. Markers constrain the flooding process only inside their own catchment basin; hence the final number of regions is equal to the number of markers.

In our semantic approach of watershed segmentation, called semantic watershed, certain regions play the role of markers/seeds. A subset of regions $S \subseteq G$ is selected to be used as seeds for the initialization of the semantic watershed algorithm. The criteria for selecting a region to become a seed, i.e. $s \in S$, are the following two:

1. The height of its fuzzy set L_s (maximum degree of membership in the fuzzy set [12]) should be above a threshold: $h(L_s) \equiv \max_{c_k \in C}(\mu_s(c_k)) > T_{seed}$. Threshold T_{seed} is calculated once in the beginning of the algorithm, based on the histogram of all degrees of membership over all regions of the image.
2. The specific region has only one dominant concept, i.e. the rest concepts should have low degrees of membership comparatively to that of the dominant concept:

$$h(L_s) > \sum_{c_k \in \{C-c^*\}} \mu_s(c_k), \text{ where } c^*: \mu_s(c^*) = h(L_s) \tag{6}$$

These two constrains ensure that the specific region has been correctly selected as seed for the particular concept c^*.

An iterative process begins checking for every initial region seed $s \in S$ in all its direct neighbors $n \in N_s$ (as defined in the ARG) if they have been assigned to the same concept c and, with what degree of membership $\mu_n(c_k)$. Some of those regions, that satisfy an additional criterion, form a new set of regions M^i (i denotes the iteration step, with $M^0 \equiv S$), which will be the new seeds for the next iteration of the algorithm. These additional criterion is that the degree of membership of region n under examination, for the particular concept c should be above a merging threshold: $\mu_n(c_k) > K^i \cdot T_{merge}$, where K is a constant slightly above one, that increases the threshold in every iteration i of the algorithm in a non linear way to the distance from the initial regions-seeds. When the above criteria are satisfied, region n is merged with its propagator s and an updated degree of membership is calculated using the default t-norm for the newly created region:

$$\mu_{\hat{s}}(c_k) = \min(\mu_s(c_k), \mu_n(c_k)) \tag{7}$$

The termination criterion of the algorithm is quite straightforward: repeat this procedure until the set of regions-seeds in step i is empty: $M^i = \varnothing$. In this point, we should underline that when neighbors of a region are examined, previous accessed regions are excluded, i.e. each region is reached only once and that is by the closest region-seed, as defined in the ARG.

After running this algorithm onto an image, some regions will be merged with one of the seeds, while other will stay unaffected. In order to deal with these regions as well, we run again the algorithm on a new ARG each time that consists of the regions that remained intact after all previous iterations. This hierarchical strategy needs no additional parameters, since every time new regions-seeds will be created automatically based on a new threshold T_{seed} (apparently with smaller value than before). Obviously, the regions created in the first pass of the algorithm have stronger confidence for their boundaries and their assigned concept than those created in a later pass. This is not a drawback of the algorithm; quite on the contrary, we consider this fuzzy outcome to be actually an advantage as we maintain all the available information.

3.3 Semantic RSST

Traditional RSST [16] is a bottom-up segmentation algorithm that begins from the pixel level and iteratively merges similar neighbor regions until certain termination criteria are satisfied. RSST is using internally a graph representation of image regions, like the ARG described in section 2.2. In the beginning, all edges of the graph are sorted according to a criterion, e.g. color dissimilarity of the two connected regions using Euclidean distance of the color components. Then recursively the edge with the least weight is found and the two regions connected by that edge are merged. After each step, the merged region's attributes (e.g. region's mean color) is re-calculated. Traditional RSST will also re-calculate weights of related edges as well and resort them, so that in every step the edge with the least weight will be selected.

Following the conventions and notation used so far, we introduce here a modified version of RSST, called Semantic RSST (S-RSST). The first step is to populate the set of edges E by traversing the ARG. In contrast to the approach described in the previous section, in this case no initial seeds are necessary, but instead of this we need to define (dis)similarity and termination criteria. The criterion for ordering the edges is based on the similaruty value defined earlier in section 2.2. Commutativity and associativity axioms of all fuzzy set operations (thus including default t-norm and default s-norm) ensure that the ordering of the arguments is indifferent. In this way all graph's edges are sorted by their weight:

$$w(e_{ab}) = 1 - s_{ab} \qquad (8)$$

Equation (8) can be expanded by substituting s_{ab} from equation (5). We considered that an edge's weight should represent the degree of dissimilarity between the two joined regions; therefore we subtract the estimated value from one.

Let us now examine in details one iteration of the S-RSST algorithm. Firstly, the edge with the least weight is selected as:

$$e_{ab}^{*} = arg \min_{e_{ab} \in E} (w(e_{ab})) \qquad (9)$$

Then regions a and b are merged to form a new region $\hat{a}$. Vertex v_b is removed completely from the ARG, whereas a is updated appropriately. This update procedure consists of the following two actions:

1. Update of the fuzzy set L_a by re-evaluating all degrees of membership in a weighted average fashion:

$$\mu_{\hat{a}}(c_k) = \frac{A(a) \cdot \mu_a(c_k) + A(b) \cdot \mu_b(c_k)}{A(a) + A(b)}, \quad \forall c_k \in C \tag{10}$$

The quantity $A(a)$ is a measure of the size of a and is the number of pixels belonging to this region.

2. Re-adjustment of the ARG's edges:
   a. Removal of edge e_{ab}.
   b. Re-evaluation of the weight of all affected edges e: the union of those incident to region a and of those incident to region b: $e \in E_a \cup E_b$.

This procedure continues until the edge e^* with the least weight in the ARG is bigger than a threshold: $w(e^*) > T_w$. This threshold is calculated in the beginning of the algorithm, based on the cumulative histogram of the weights of E.

4 Visual Context

The idea behind the use of visual context information responds to the fact that not all human acts are relevant in all situations and this holds also when dealing with image analysis problems. Since visual context is a difficult notion to grasp and capture [17], we restrict it herein to the notion of ontological context. The latter is defined within the "fuzzified" version of traditional ontologies presented in section 2.1 and the problems to be addressed include how to meaningfully readjust the membership degrees of the merged regions after the semantic region growing algorithm application and how to use visual context to influence the overall results of knowledge-assisted image analysis towards its best performance.

Based on the mathematical foundations described in previous subsections, we introduce the algorithm used to readjust the degree of membership $\mu_a(c_k)$ of each concept $c_k \in C$ associated to a region $a \in G$ of the scene. Each specific concept c_k is present in the application-domain's ontology, stored together with its relationship degrees r_{c_k, c_j} to any other related concept c_j. To tackle cases that more than one concept is related to multiple concepts, the term *context relevance* $cr_{dm}(c_k)$ is introduced, which refers to the overall relevance of concept c_k to the *root element* characterizing each domain dm. For instance the root element of beach and motorsports domains are concepts c_{beach} and $c_{motorsport}$ respectively. All possible routes in the graph are taken into consideration forming an exhaustive approach to the domain, with respect to the fact that all routes between concepts are reciprocal.

Estimation of each concept's value is derived from direct and indirect relationships of the concept with other concepts, using a meaningful *compatibility indicator* or distance metric. Depending on the nature of the domains under consideration, the best indicator could be selected using the *max* or the *min* operator, respectively. Of course the ideal distance metric for two concepts is one that quantifies their semantic correlation. For the problem at hand and given both the beach and motorsports domains, the *max* value is a meaningful measure of correlation. A simplified example, limiting the only available concepts to $c_{motorsport} = c_m$, $c_{asphalt} = c_a$, $c_{grass} = c_g$ and $c_{car} = c_c$ is presented in Fig. 1 and summarized in the following: letting concept c_a be related to concepts c_m, c_g and c_c directly with: r_{c_a,c_m}, r_{c_a,c_g} and r_{c_a,c_c}, while concept c_g is related to concept c_m with r_{c_g,c_m} and concept c_c is related to concept c_m with r_{c_c,c_m}. Additionally, c_c is related to c_g with r_{c_c,c_g}. Then, we calculate the value for $cr_{dm}(c_a)$:

$$cr_{dm}(c_a) = \max \left\{ \begin{matrix} r_{c_a,c_m}, \; r_{c_a,c_g} \cdot r_{c_g,c_m}, \; r_{c_a,c_c} \cdot r_{c_c,c_m}, \\ r_{c_a,c_g} \cdot r_{c_g,c_c} \cdot r_{c_c,c_m}, \; r_{c_a,c_c} \cdot r_{c_c,c_g} \cdot r_{c_g,c_m} \end{matrix} \right\} \tag{11}$$

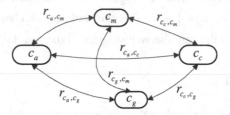

Fig. 1. Graph representation example – Compatibility indicator estimation

The general structure of the proposed re-evaluation algorithm is summarized in the following steps:

1. Identify an optimal normalization parameter np to use within the algorithm's steps, according to the considered domain(s). The np is also referred to as domain similarity, or dissimilarity, measure and $np \rightarrow [0,1]$.

2. For each concept $c_k \in C$ in the fuzzy set L_a associated to a region $a \in G$ in a scene with a degree of membership $\mu_a(c_k)$, obtain the particular contextual information in the form of its relations to the set of any other concepts: $\{r_{c_k,c_j} : c_j \in C, \; c_j \neq c_k\}$.

3. Calculate the new degree of membership $\mu_a(c_k)$ associated to region a, based on np and the context's relevance value. In the case of multiple concept relations in the ontology, relating concept c_k to more than one concepts, rather than relating c_k solely to the "root element" c_r, an intermediate aggregation step should be applied for c_k: $cr_{c_k} = \max\{r_{c_k,c_r}, ..., r_{c_k,c_m}\}$. We express the calculation of $\mu_a(c_k)$ with the recursive formula:

$$\mu_a^i(c_k) = \mu_a^{i-1}(c_k) - np \cdot (\mu_a^{i-1}(c_k) - cr_{c_k}) \tag{12}$$

Where i denotes the iteration used. Equivalently, for an arbitrary iteration i:

$$\mu_a^i(c_k) = (1-np)^t \cdot \mu_a^0(c_k) + (1-(1-np)^i)cr_{c_k} \qquad (13)$$

Where $\mu_a^0(c_k)$ represents the original degree of membership. Typical values for i reside between 3 and 5.

Key point in this approach remains the definition of a meaningful normalization parameter np. When re-evaluating this value, the ideal np is always defined with respect to the particular domain of knowledge and is the one that quantifies each semantic correlation to the domain. In this work we conducted a series of experiments on a "training" subset of 52 images for both application domains and selected the np that resulted in the best overall precision/recall values for each domain.

5 Experimental Results

We carried out experiments in the domains of beach and motorsports, utilizing 262 images in total, i.e. 193 beach and 69 motorsports images acquired either from the internet or from personal collections. In order to demonstrate the proposed methodologies and keep track of each individual algorithm results, we integrated the described techniques into a single application that utilizes a user-friendly graphical interface. In the following we present two representative sets of experimental results, i.e. one image derived from the beach domain and one image from the motorsports domain. Each set includes four images: (a) the original image, (b) the result of traditional RSST, (c) the result of semantic watershed and (d) the result of semantic RSST. In the case of the traditional RSST, we pre-defined the final number of regions to be produced to be equal to the ones produced by the semantic watershed; in this fashion segmentation results are easily comparable.

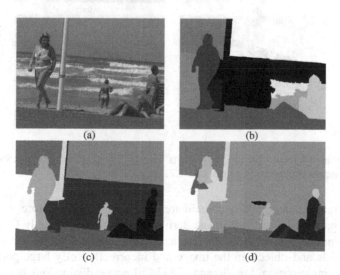

Fig. 2. Experimental results for the beach domain – Example 2. (a) Input image, (b) RSST segmentation, (c) semantic watershed, (d) semantic RSST.

Fig. 2 illustrates the example derived from the beach domain. As observed in Fig. 2b, RSST segmentation results are insufficient: some persons are merged with sea segments, while others are not detected at all and most sea regions are divided because of the waves. Semantic watershed application results into significant improvements (Fig. 2c). Sea regions on the left part of the image are successfully merged together, the woman on the left is correctly identified as one region, successfully tackling the existence of variations in low level characteristics, i.e. green swimsuit vs. color of the skin, etc. Persons on the right side are identified and not merged with sea or sand regions, having as a side effect the fact that there are multiple persons in the image and not just a single one. Very good results are obtained in the case of the sea in the right region, although it is inhomogeneous in terms of color and material because of the waving. We observe that it is successfully merged into one region and the person standing in the foreground is also identified as a whole. Finally, semantic RSST algorithm in Fig. 2d performs similarly well. Small differences with semantic watershed are justified by the fact that in S-RSST focus is given on material and not in objects in the image. Consequently, persons are identified with greater accuracy in the image and are segmented, but not wrongly merged, e.g. the woman on the left is composed by multiple regions due to the nature of the material or people on the right are composed by different regions.

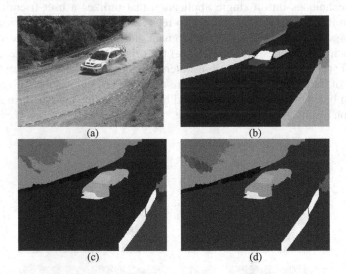

(a)   (b)

(c)   (d)

Fig. 3. Experimental results for the motorsports domain. (a) Input image, (b) RSST segmentation, (c) semantic watershed, (d) semantic RSST.

Results from the motorsports domain are described in Fig. 3. More specifically, in Fig. 3a we present the original image derived from the World Rally Championship. Plain segmentation results (Fig. 3b) are again poor, since they do not identify correctly materials and objects in the image and incorrectly unify large portions of the latter into a single region. Fig. 3c and Fig. 3d illustrate distinctions between vegetation and cliff regions in the upper left corner of the image. Even different vegetation areas are identified as different regions in the same area. Furthermore, the car's wind-shield remains correctly a standalone region, because of its large color and material diversities in comparison to the regions in its neighborhood. Because of the difficulties and obstacles set by the nature of the image, the thick shadow in the front of the car is inevitably unified with the front dark part of the latter and the "gravel smoke" on the side is recognized as gravel, resulting into a deformation of the vehicle's chassis. These are two cases where both semantic region growing algorithms seem to perform poorly. This is due to the fact that the corresponding segments differ visually and the possible detected object is a composite one - in contradiction to the so far encountered material objects - and is composed by regions of completely different characteristics. Furthermore, on the right side of the image, the yellow ribbon is dividing two similar but not identical gravel regions, fact that is correctly identified by our algorithm. The main difference between the SW and SRSST approaches is summarized in the way they handle vegetation in the upper left corner of the image, with SRSST performing closer to the ground truth, since it detects the variations in vegetation and grass successfully.

Finally, we continue with presenting a visualization of the contextualization step implemented within our approach. In general, our context algorithm successfully aids in the determination of regions in the image and corrects misleading behaviors, originating from over- or under-segmentation, by meaningfully adjusting confidence values.

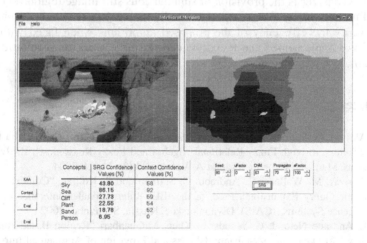

Fig. 4. Contextual experimental results for the first beach image example

In Fig. 4 we observe the contextualization step for the first beach image, presented in the Contextual Analysis tool developed. Contextualization, which works on a per region basis, is applied after the semantic region merging, in order for its results to be meaningful. We have selected the unified sea region in the upper left part of the image, as illustrated by its blue color. The contextualized results are presented in red in the right text column at the bottom of the GUI. Context favors strongly the fact that the merged region belongs to sea, increasing its confidence value from 86.15% to a crisp 92%. Additionally, the totally irrelevant (for the specific region) confidence value for the concept person is extinguished, whereas medium confidence values for the rest of the possible beach concepts are slightly increased, due to the ontological knowledge relations that exist in the considered knowledge model. That is because of the relationships that exist in the a priori built contextual knowledge and that strongly relate concepts encountered on a beach scene with each other, we expect that the use of context will improve the results but at the same time provide also some false concepts as well. However, in all cases context does normalize results in a meaningful manner, i.e. each region's dominant concept is detected in comparison to ground truth and its degree of membership is increased.

6 Conclusion

The methodologies presented in this paper can be exploited towards the development of more intelligent and efficient image analysis environments. Image segmentation and detection of objects based on the semantic level, with the aid of contextual information, results into meaningful results. The core contributions of the overall approach have been the implementation of two novel semantic region growing algorithms, acting independently from each other, as well as a novel visual context interpretation based on an ontological representation, exploited towards optimization of region' associated fuzzy set of concepts provided by the segmentation results. Another important point to consider is the provision of simultaneous still image region segmentation and labeling, providing a new aspect to traditional object detection techniques. In order to verify the efficiency of the proposed algorithms when faced with real-life data, we have implemented and tested them in the framework of developed research applications.

References

[1] G. Akrivas, G. Stamou and S. Kollias, "Semantic Association of Multimedia Document Descriptions through Fuzzy Relational Algebra and Fuzzy Reasoning", IEEE Trans. on Systems, Man, and Cybernetics, part A, Volume 34 (2), March 2004

[2] G. Akrivas, M. Wallace, G. Andreou, G. Stamou and S. Kollias, "Context – Sensitive Semantic Query Expansion", Proc. of the IEEE International Conference on Artificial Intelligence Systems (ICAIS), Divnomorskoe, Russia, September 2002

[3] E. L. Andrade Neto, J. C. Woods, E. Khan, M.Ghanbari "Region Based Analysis and Retrieval for Tracking of Semantic Objects and Provision of Augmented Information in Interactive Sport Scenes" IEEE Trans. on Multimedia Vol. 7, Issue 6, pp. 1084–1096, December 2005

[4] Th. Athanasiadis, V. Tzouvaras, K. Petridis, F. Precioso, Y. Avrithis and Y. Kompatsiaris, "Using a Multimedia Ontology Infrastructure for Semantic Annotation of Multimedia Content", Proc. of 5th International Workshop on Knowledge Markup and Semantic Annotation (SemAnnot '05), Galway, Ireland, November 2005

[5] Th. Athanasiadis, Y. Avrithis, S. Kollias, "A Semantic Region Growing Approach in Image Segmentation and Annotation", 1st International Workshop on Semantic Web Annotations for Multimedia (SWAMM), Edinburgh, Scotland, November 2006

[6]   A. B. Benitez, H. Rising, C. Jrgensen, R. Leonardi, A. Bugatti, K. Hasida, R. Mehrotra, M. Tekalp, A. Ekin, and T. Walker, "Semantics of Multimedia in MPEG-7", In IEEE International Conference on Image Processing, pages 137--140, vol.1, 2002

[7]   S. Beucher and F. Meyer, "The Morphological Approach to Segmentation: The Watershed Transformation", Mathematical Morphology in Image Processing, E.R.Doughertty (Ed.), Marcel Dekker, NY, 1993

[8]   E. Borenstein, E. Sharon and S. Ullman, "Combining Top-Down and Bottom-Up Segmentation", Computer Vision and Pattern Recognition Workshop, Washington DC, USA, June 2004

[9]   M. Boutell, J. Luo, X. Shena and C. Brown, "Learning multi-label scene classification", Pattern Recognition, 37(9), pp. 1757-1771, September 2004

[10] T.R. Gruber, "A Translation Approach to Portable Ontology Specification", Knowledge Acquisition 5: 199-220, 1993

[11] F. Jianping, D. K. Y. Yau, A. K. Elmagarmid and W. G. Aref, "Automatic image segmentation by integrating color-edge extraction and seeded region growing", IEEE Trans. on Image Processing, Vol. 10, No. 10, pp. 1454-1466, October 2001

[12] G. Klir and B. Yuan, "Fuzzy Sets and Fuzzy Logic, Theory and Applications", New Jersey, Prentice Hall, 1995

[13] S. Lee, M. M. Crawford, "Unsupervised classification using spatial region growing segmentation and fuzzy training", Proc. of the IEEE International Conference on Image Processing, Thessaloniki, Greece, 2001

[14] J. Luo and A. Savakis, "Indoor vs outdoor classification of consumer photographs using low-level and semantic features", In Proc. IEEE Int. Conf. on Image Processing (ICIP01), 2001

[15] B.S. Manjunath, J.R. Ohm, V.V. Vasudevan, A. Yamada, "Color and texture descriptors", Special Issue on MPEG-7, IEEE Trans. on Circuits and Systems for Video Technology, 11/6, 703-715, June 2001

[16] O.J. Morris, M.J. Lee and A.G. Constantinides, "Graph theory for image analysis: An approach based on the shortest spanning tree", Proc. Inst. Elect. Eng., vol. 133, April 1986, pp. 146–152

[17] Ph. Mylonas and Y. Avrithis, "Context modeling for multimedia analysis and use", Proc. of 5th Inter-national and Interdisciplinary Conference on Modeling and Using Context (CONTEXT '05), Paris, France, July 2005

[18] P. Salembier, F. Marques, "Region-Based Representations of Image and Video - Segmentation Tools for Multimedia Services", IEEE Trans. on Circuits and Systems for Video Technology, vol.9, no.8, 1999

[19] T. Sikora, "The MPEG-7 Visual standard for content description - an overview", Special Issue on MPEG-7, IEEE Trans. on Circuits and Systems for Video Technology, 11/6:696-702, June 2001

[20] ISO/IEC JTC 1/SC 29/WG 11/N3966, "Text of 15938-5 FCD Information Technology – Multimedia Content Description Interface – Part 5 Multimedia Description Schemes", Singapore, 2001

[21] W3C, RDF Reification, http://www.w3.org/TR/rdf-schema/#ch_reificationvocab

Automated Speech and Audio Analysis for Semantic Access to Multimedia

Franciska de Jong[1,2], Roeland Ordelman[1], and Marijn Huijbregts[1]

[1] University of Twente, Dept. of Computer Science,
P.O. Box 217, 7500 AE, Enschede, The Netherlands
[2] TNO-ICT, Delft, The Netherlands
{fdejong, ordelman, m.a.h.huijbregts}@ewi.utwente.nl
http://hmi.ewi.utwente.nl

Abstract. The deployment and integration of audio processing tools can enhance the semantic annotation of multimedia content, and as a consequence, improve the effectiveness of conceptual access tools. This paper overviews the various ways in which automatic speech and audio analysis can contribute to increased granularity of automatically extracted metadata. A number of techniques will be presented, including the alignment of speech and text resources, large vocabulary speech recognition, key word spotting and speaker classification. The applicability of techniques will be discussed from a media crossing perspective. The added value of the techniques and their potential contribution to the content value chain will be illustrated by the description of two (complementary) demonstrators for browsing broadcast news archives.

1 Introduction

The growing role expected for networked electronic media and the increasing size of content repositories require augmented attention for the automation of content-based extraction and integration of metadata for video, audio and textual content. Content-based metadata are a prerequisite for conceptual search both for professional users and the general public, and they play an important role in the exploitability of content. The adagium used to be 'Content is king', but metadata rules.

Semi-automatic metadata extraction has been given attention for a variety of monomedia content types and formats. For video and image content this has led to an interesting growth in the number of objects and events that can be detected on the basis of low-level features [13]. But in view of the huge range of concepts that users may want to search for, the field of video analysis can be argued to be still in its infancy.

A strategy that can help compensate for the limitations of image analysis is the exploitation of surface features. Surface features are those properties of (multimedia) documents that do not describe content. Examples include the length of a document, references to the document's location, and the production date. Although these features do not directly relate to the document's coverage,

Y. Avrithis et al. (Eds.): SAMT 2006, LNCS 4306, pp. 226–240, 2006.
© Springer-Verlag Berlin Heidelberg 2006

they have proven to be valuable additional sources of information in a retrieval setting [17]. More importantly they illustrate the importance of not being too restrictive in exploiting available secondary data streams.

As is widely acknowledged, the exploitation of linguistic content in multimedia archives can boost the accessibility of multimedia archives enormously. Already in 1995, [3] demonstrated the use of subtitling information for retrieval of broadcast news videos, and in the context of TRECVID [13] the best performing video retrieval systems always exploit speech transcripts. The added value of linguistic data is of course limited to video data containing textual and/or spoken content, or to video content with links to related textual documents, e.g. subtitles, generated transcripts etc. But when available, using linguistic content for the generation of a time-coded index can help to bridge the semantic gap between media features and user needs.

In the next two sections we first explore some methods that deal with the exploitation of already available linguistic content in, or attached to, multimedia databases. We introduce the concept of cross-media mining in section 4. Automatic audio indexing techniques are overviewed in section 5. The system architecture of the recognition environment is detailed in section 6. Finally, the added value of links that are automatically generated across media via high level annotation will be illustrated in section 7. This section will provide a description of two complementary demonstrators: one for on-line access to an archive of news broadcasts linked up to a newspaper archive, the other illustrating a crucial aspect in browsing multimedia databases, a technique known as *document clustering* applied in combination with topic detection.

2   Exploiting Collateral Text

The semantic gap between user needs and content features is as old as the concept of archiving itself. The traditional approach towards the creation of an index is to rely on manual annotation with controlled vocabulary index terms. With the emergence of digital archiving this approach is still widely in use and for many archiving institutes the creation of manually generated metadata is and will be an important part of the daily work. When the automation of metadata generation is considered, it is often seen as something that can enhance the existing process rather than replace it. The available metadata will therefore often be a combination of highly reliable and conceptually rich annotations, and (semi)automatically generated metadata. One of the challenges for search environments is to combine the various types of metadata and to exploit the added value of the combination. In this paper we will explain how available high level annotations for media archives can be exploited for improved automated generation of additional language-based annotations, and *vice versa*, how automatic content processing can help to generate ontological and thesaurial media annotations. For the content-based processing tasks the main focus will be on the various ways in which automatic speech and audio analysis can be deployed.

Depending on the resources available within an organization that administers a media collection, the amount of detail of the metadata and their characteristics may vary. Large national audiovisual institutions such as Beeld&Geluid in The Netherlands[1], annotate at least titles, dates and short content descriptions (descriptive metadata). Many organizations with multimedia collections however, often do not have the resources to apply even some basic form of archiving.

To still allow the conceptual querying of video content, collateral textual resources that are closely related with the collection items can be exploited. A well known example of such a textual resource is subtitling information for the hearing-impaired (e.g., CEEFAX pages 888 in the UK) that is available for the majority of contemporary broadcast items, in any case for news programs. Subtitles contain a nearly complete transcription of the words spoken in the video items and provide an excellent information source for indexing. Usually, they can easily be linked to the video by using the time-codes that come with the subtitles. The Dutch news subtitles even provide topic boundaries that can be used for segmenting the news show into subdocuments. Textual sources that can play a similar role are teleprompter files: the texts read from screen by an anchor person (also referred to as auto-cues).

The time labels in these sources are crucial for the creation of a textual index into video. As in full text retrieval, where all words in a document can function as index terms and thus as a link to a document, the exploitation of collateral transcriptions for speech in video will allow that all words spoken offer a link to the fragment in which they occur. And though full text retrieval is certainly not the ultimate solution to the semantic gap, natural language is inherently closer to the level of concepts than low-level image features.

3   Time Alignment

In the collateral text sources mentioned above, the available time-labels are not always fully reliable and can even be absent. In such cases the text files will have to be synchronized. Examples of such text sources are minutes of meetings, or written versions of lectures and speeches. This section will describe methods for the automatic generation of time-stamps for minutes in two pilot projects in the domain of e-Government. These minutes pertain to the so-called *Handelingen*, i.e. the meetings of the Dutch Parliament, and to city council meetings. Due to the difference in accuracy of the minutes, two different approaches had to be developed.

The minutes of the meetings of the Dutch Parliament are stenographic minutes that closely follow the discourse of the meeting, only correcting slips of the tongue and ungrammatical sentences. Given the close match with the actual speech, a relatively straightforward so-called forced alignment procedure could be used. Forced alignment is a technique commonly used in acoustic model training in automatic speech recognition (ASR). In order to be able to train phone models, words and phones in pre-segmented sentences are aligned to their exact location

---

[1] Beeld&Geluid:http://www.beeldengeluid.nl/

in the speech segments using an acoustic model[2]. Given a set of words from a sentence the acoustic model tries to find the most optimal distributions of these words given the audio signal on the basis of the sounds the words are composed of. When using alignment for indexing, pre-segmented sentences are evidently not available but as long as the text follows the speech well enough, the word alignment can be found by using relatively large windows of text.

The alignment procedure works well even if some words in the minutes are actually not in the speech signal. However, if the text to be aligned does not match the speech too well, as was the case with city council meetings, and if the text segments are too large, the alignment procedure will fail to find a proper alignment. In order to produce suitable segments, we used a two-pass strategy, similar as proposed in [9], incorporating the following steps:

1. a baseline large vocabulary speech recognition system[3] is used to generate a relatively inaccurate transcript of the speech with word-timing;
2. the transcript is aligned on the word level to the minutes using a dynamic programming algorithm;
3. where the transcript and the minutes match (three words in a row correctly aligned), so-called 'anchors' are inserted
4. using the word-timing labels provided by the speech recognition system, the anchors are used to generate suitable segments;
5. individual segments of audio and text are accurately synchronized using forced alignment;

The described methods allow for the synchronization of audiovisual data to available linguistic content that approximates to a certain extent the speech in the source data and they enable the processing of conceptual queries of the audiovisual content with readily available tools.

4   Cross-Media Mining

Ideally one would not only synchronize audiovisual material with content that approximates the speech in the data, but take even one step further and exploit *any* collateral textual resource, or even better: any kind of textual resource that is accessible, including open source titles and proprietary data (e.g., trusted webpages and newspaper articles). Another way of putting it: we propose to shift the focus from indexing individual multimedia documents to video mining in truly multimedia distributed databases. In the context of meetings for example, usually an agenda, documents on agenda topics and cv's of meeting participants can be obtained and added to the repository. Mining these resources can support information search because it yields annotations that offer the user not just access to a specific media type, but also different perspectives on the available

---

[2] In the first iteration usually an 'averaged' bootstrap model is used. The alignment and the model should improve iteratively.
[3] Optionally the speech recognition is somewhat adapted to the task for example by providing it with a vocabulary extracted from the minutes.

data. An agenda could help to add structure that can for example be presented
in a network representation, whereas cv's can be linked to annotations resulting
from automatic speaker segmentation. In addition, both documents and cv's
would allow for multi-source information extraction.

A typical example of what the cross-media perspective can yield in the broadcast news domain is the linking of newspaper articles with broadcast items and
vice versa (Cf. Fig 1). Links can be established between two news objects which
count as similar on the basis of the language models assigned to them via statistical analysis. Typically such language models are determined by the frequency
of the linguistic units such as written or spoken words and their co-occurrences.
The similarity between two documents can be decided for each pair of documents, but a more common approach is to pre-structure a document collection
into clusters of documents with similar language models. Similarity of language
models predicts similarity of topic, and therefore this technique is know as topic
clustering[4].

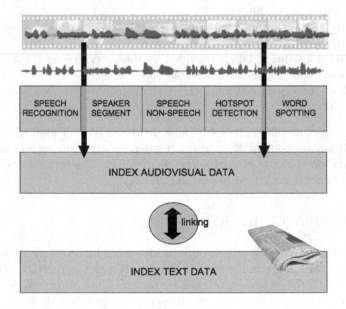

Fig. 1. Linking audio and textual sources

In addition to linking documents with a similar topic profile, which can be
supportive in a browser environment, also the available semantic annotation for
documents with similar profiles can be exchanged and exploited for conceptual
search. If a newspaper article has been manually classified as belonging to e.g.,
economy or foreign politics, a broadcast item with a similar language model can

---

[4] The functionality commonly known as *topic detection and tracking* (TDT) for dynamic news streams has been built upon it and plays a central role in the evaluation
series for TDT organized by DARPA.

be classified with these conceptual labels as well. In section 7 below, we describe a cross-media news browser demonstrator that incorporates this functionality.

For the linking of audiovisual data with textual resources that are not directly related to the speech, such as documents on agenda topics in the context of meetings, or newspaper data in the broadcast news domain, we have to step up the use of speech recognition compared to the speech recognition deployed in the alignment procedures described in section 3. In the more elaborate alignment procedure, an initial hypothesis is generated by a large vocabulary speech recognition system. As this hypothesis is only needed for finding useful segments, we do not really care about the performance of the system as long as it is able to provide us with 'anchors'. However, the relevance of speech recognition performance increases when textual resources suitable for alignment with audiovisual data are *not* available. In the next section, the application of speech recognition technology as the *primary* source for generating a textual representation of audiovisual documents that can be linked to other linguistic content, is described.

5   Audio Indexing

Recent investigations have shown the feasibility of deploying large vocabulary speech recognition for the generation of multimedia annotations that allow the conceptual querying of video content and the synchronization to any kind of textual resource that is accessible, including other full-text annotation for audiovisual material. The potential of ASR-based indexing has been demonstrated most successfully in the broadcast news domain. Data collection for training a speech recognition system in this general domain is relatively easy. Word-error-rates below 10% are no longer exceptional. For the broadcast news domain, ASR transcripts approximate the quality of manual transcripts, at least for several languages. Spoken document retrieval in the American-English broadcast news (BN) domain has even been declared 'a solved problem' with the NIST-sponsored TREC SDR track in 1999 [6]. It should be noted however that in other domains than broadcast news and for many less favored languages, a similar recognition performance is usually harder to obtain due to (i) lack of domain-specific training data, and (ii) large variability in audio quality, speech characteristics and topics being addressed. However, when recognition performance remains within certain boundaries (an ASR performance of 50% WER is typically regarded as a lower bound for successful retrieval) the damage in terms of retrieval performance may be acceptable, especially when no other means (metadata) are available for searching.

5.1   Vocabulary Selection Via Collateral Text

One of the main research topics in large vocabulary speech recognition is vocabulary selection. Given the huge quantities of available training data for the broadcast news domain, the acoustic models and language models can usually be trained adequately and in addition, various acoustic adaptation procedures (using e.g., bandwidth/gender-dependent models, speaker-adaptive training, etc.)

can be applied to boost ASR performance. However, language models and recognition vocabularies are usually created using fixed and, with respect to broadcast news, often outdated training corpora. Vocabularies are based on word frequencies within corpora, while the linguistic properties in broadcast news are continuously changing: previously infrequent names of places and people can start occurring frequently without prior indication, people dominating the news during a period of time may disappear from the headlines after a while, jargon may suddenly be adopted by the general public, new words are invented and there are words that are likely to (co-)occur in one period of the year but highly improbable in another period (e.g., *hurricane*, or *Christmas tree*.

To limit the number of out-of-vocabulary (OOV) words, the ASR engine of an SDR environment should be based on models that adapt to linguistic variation. OOVs damage retrieval performance in two ways: firstly, a query consisting of an OOV word, a so-called QOV (query-out-of-vocabulary), will never match a transcript, even though the QOV occurs in the speech. Secondly, the word occurring in the transcript at the position of the OOV may induces a false hit. Although document expansion and query expansion techniques may be deployed to compensate for QOV words [18,7], tackling OOVs in an earlier stage is favorable. As named entities play an important role in the mining of broadcast news, regular updates of a recognition vocabulary with 'new' proper names is crucial. To keep an ASR engine 'tuned', up-to-date training material is required. Ideally, this is dealt with via a daily feed of newspaper content. Alternatively contemporary data can be collected via the Web [1] or by capturing subtitling information from news programs. A number of vocabulary selection methods have been proposed, based on parallel corpora. They are based on the use of a narrow look-back time window to select new words [2], the use of word history information [11], or the use of vectors combining word frequency data from multiple corpora [1].

5.2 Word Spotting

Another way of reducing the effect of OOV words is to use word spotting. Word spotting is the audio search functionality that matches query terms for audio content either directly or via a phone (or phone-lattice) representation of query and content (cf. [8]) and can be very effective, especially when the vocabulary for the domain is hard to predict, resulting in high OOV-rates. Word spotting can be combined with ASR based on a full text transcription to 'correct' OOVs or misrecognised names. In this approach the following steps are taken: (i) the initial ASR transcript is used to identify related collateral textual data; (ii) named entity detection applied to the collateral textual data sources provides a list of named entities that are relevant given the audio document topic; (iii) the occurrence (and timestamps) of these named entities in the audio are recovered using a word spotting approach.

5.3 Speaker Classification

There is more information in the speech than the words alone. Speaker characteristics can be extracted from the speech (speaker's voice, word usage, syntax) as well and may serve as an additional information layer, for example to add structure for browsing (speaker segmentation and identification) or to extract features that could not be accessed using traditional views on the data. Automatic speaker classification can especially be beneficial for spoken documents in cultural heritage collections. Historical spoken word archives receive attention from professional information analysts in various fields. Historians for example, may be interested both in the exact words that were spoken, but also in the speaker's profile. The latter may partly be reconstructed using the identification of speaker characteristics such as accent, age, gender, speaking behavior and even emotion and cognitive state.

Instead of aiming at the extraction of speech features from a single speaker's voice, research has been directed to extract features from multiple voices, for example emotional features in order to detect so called 'hot spots' in collections. Typical examples of such 'hot spot' detections are the cheering of a crowd in a sports game, or laughter in the context of meetings (cf. [15]).

6   System Description

In this section the ASR system architecture will be discussed in more detail. As indicated above, ASR can be deployed in different scenarios and for content in several domains, both within and outside the broadcast domain: documentaries, interviews, historical archives, recordings of lectures and recordings of meetings (corporate, scientific, parliamentary, etc.). As the characteristics of the speech encountered in these domains vary, robustness of recognition is taken as a requirement. A robust speech recognition system can be defined as a system that is capable of maintaining good recognition performance even when the quality of the speech input is low (environment, background noises, cross-talk, low audio quality), or when the acoustical, articulatory, or phonetic characteristics of the speech encountered in the training data differ from the speech in the task data. Even systems that are designed to be speaker independent cannot cover all the speaker variations that may occur in the different task domains.

As manually adjusting a system for each different situation is time consuming and error prone, we designed and developed an ASR system that is self-adjustable as much as possible, and that is robust for changing conditions. In Figure 2 the system is graphically represented. The gray area at the left represents the sub-system responsible for adapting our baseline, broadcast news (BN), knowledge models to a specific domain. This sub-system will be described in section 6.1. The second gray area represents the main audio processing stream. This sub-system consists of four modules. In the first module, the audio is cut up in smaller segments and each segment is assigned to a unique speaker. This process of determining 'who spoke when', called speaker diarization, will be discussed in section 6.2. After diarization, the second module performs a first speech recognition pass for each speaker using the domain specific models. This recognition result is then used in the third module to create a new acoustic model (AM) for each speaker. These speaker specific models are eventually used in the fourth module doing the final decoding pass. The decoding procedure is discussed in section 6.3. Some first attempts were made to make the system more robust by monitoring and analyzing the system's performance. In section 6.4 some of our monitoring tools are discussed.

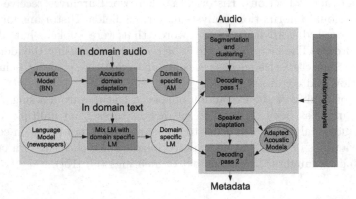

Fig. 2. The ASR system consisting of four modules: first the audio is divided in segments and each segment is assigned to a speaker. The speech of each speaker is decoded and used for adapting the acoustic model. The adapted models are used in the second decoding pass.

6.1   Acoustic and Language Model Creation

The system uses two kinds of knowledge models. The acoustic model (AM) is used for calculating which sequence of phones has the highest probability of being pronounced at a certain time in the audio stream. The language model (LM) is used for calculating which sequence of words is most likely pronounced. Large amounts of training data are required to extract the statistics that are needed to create these models. Unfortunately, for most application domains outside the news domain, data is not sufficiently available for model training.

For the broadcast news domain relatively large text corpora are available. The Twente News Corpus contains over 450M words of newspaper text data that are used to train our BN language model. The BN acoustic model is trained on approximately 150 hours of speech from a variety of sources including the Spoken Dutch Corpus [10]. (Partly) unsupervised methods to augment the basis for training could offer compensation. In order to use the system in other domains than broadcast news, we typically use small amounts of in-domain audio and text data to adapt our BN models to the new task domain. The goal is to adjust the acoustic model parameters so that the model better fits the task domain using a model-space transformation method (such as SMAPLR [12]).

6.2   Speaker Diarization

The task of the first module in the system is to cut the audio in smaller *segments* suitable for input to our decoder. Speech and non-speech segments are distinguished (Speech Activity Detection) and speaker changes are detected. Simultaneously the module determines 'who spoke when' and clusters the segments of unique speakers. This procedure is often called *speaker diarization*.

The speaker diarization module was evaluated in the speaker diarization task of the NIST Rich Transcription 2006 Spring Evaluation (RT06s). After all non-speech, such as silence and background noise, has been removed from the audio, the remaining segments are passed to a modified Hidden Markov Model (HMM) speech decoder. The HMM topology consists of a number of parallel single state HMMs connected to each other by a single (non-emitting) start- and end-state. Each state represents a single speaker from the audio. Because the exact number of speakers is unknown, the system will initially contain more states than the maximum expected number of speakers. After training the HMM with the input audio, the number of states is reduced until an optimum system likelihood is reached. At RT06s we have successfully tested two methods for reducing the number of states [16]. In the ideal situation, after this procedure there will be exactly as many states as there are speakers in the audio and each state will be trained on a single unique speaker. Performing a Viterbi alignment using the final HMM topology will result in a set of speech segments for each speaker.

The major advantage of the used diarization method is that hardly any parameters need to be tuned for different audio or domain conditions. As a consequence the resulting system is robust to changing conditions.

The diarization module can cluster all speech from a single speaker within a single audio document. This in-document speaker information can already be valuable metadata in itself (see section 5.3), but in order to track speakers across documents we would like to cluster speakers over document boundaries. This should be possible by extending this technology so that speaker models of different documents can be compared.

6.3   Two-Pass Decoding

Once the audio has been segmented and clustered, the speech is recognized in either one or two decoding passes. Our most recent systems use the recognition of the first pass to adapt the acoustic model to the speakers encountered in the data. This so called speaker adaptation is performed using audio with high confidence scores using the SMAPLR adaptation method [12]. The new speaker specific acoustic models are then used in the second decoding pass to produce the final recognition.

A two-stage recognition run allows for domain adaptation on the word level as well. Here, the strategy is to assign a topic to segments of the input file on the basis of the speech transcript of the first run. The topic assignment is then used to select a topic-specific language model. As evidently real topic-based segmentations are not known a-priori, readily available segmentations, such as on the change of speaker, on longer silence intervals or even fixed time-windows, are chosen to divide the audio document in smaller parts. These parts are then further regarded as representing single topics.

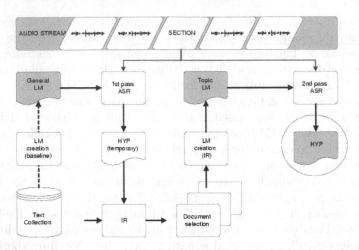

Fig. 3. Topic based LM: segmentation of the audio file, initial speech recognition on the audio segments, defining the 'topic' on the basis of the speech transcripts (here using an IR system), creating a topic specific language model, and finally the final speech recognition run using the topic-based language model.

The topics can be assigned either implicitly or explicitly. An explicit topic assignment refers to using specific topic labels, for example generated on the basis of a topic-classification system that assigns thesaurus terms. From a collateral text corpus (e.g., a newspaper corpus) that is labeled with the same thesaurus terms, documents that are similar to the topic in the segment can be harvested for creating a topic-specific language model. For implicit topic assignment, an Information Retrieval system is used for the selection of documents from an unstructured collateral text source (e.g., internet sources) that have a similar topic: on the basis of the stopped speech transcript that serves as a query, a ranked list of similar documents is generated; the top N documents of the list in turn serve as input for language modeling. Having created a topic specific language model a second speech recognition run is performed on the same segment with the new language model to generate the final transcript. The procedure is visually depicted in Figure 3.

Both recognition passes are performed with the University of Twente 2006 (UT06) decoder. The UT06 decoder is a Large Vocabulary Continuous Speech Recognition decoder that uses Hidden Markov Models (HMM). Its state emission probabilities are calculated by Gaussian Mixture Models (GMM). We have trained models based on Perceptual Minimum Variance Distortionless Response (PMVDR) cepstral coëfficient features (created using the Sonic LVCSR toolkit [19]) and with Mel Frequency Cepstral Coëfficient (MFCC) features.

6.4 System Monitoring and Output Analysis

Given that (i) a large amount of system parameters need to be fine-tuned for every application domain, and (ii) the system's behavior often needs to be monitored over a longer period of time (e.g., in longitudinal tasks), various methods to monitor and analyze the system are being developed.

One of these methods is blame assignment [4]. This method uses a small evaluation set (audio that has been manually annotated on the word level) to evaluate recognition accuracy. Incorrectly decoded words are grouped in *error regions*. For each region it is calculated in which error class the region belongs. There are five error classes: (i) the region contains out of vocabulary words, (ii) the error is caused by a LM mismatch, (iii) an AM mismatch or (iv) a mismatch in both the LM and AM, and (v) the error is caused by pruning away the correct path during decoding (search error).

The blame assignment method can only be used when evaluation data is available. When monitoring a BN system that needs to decode broadcasts on a daily basis, transcribed data may not be available. We are currently investigating system behavior over a longer period of time using collateral data sources such as subtitling information that comes with broadcast news programmes or minutes of meetings as evaluation data. The difficulty here is how the mismatch between the noisy speech transcripts containing errors and the incomplete and/or reformatted collateral text data should be interpreted.

7   The Power of Transcripts Demonstrated

A number of techniques described above have been implemented in two demonstrators described below. They illustrate how, on the basis of textual transcripts, the concept of cross-media news browsing for a multifaceted or layered multimedia archive can be realized.

7.1   Cross-Media News Browser

The so-called cross-media news browser demonstrates on-line access tools to an archive of Dutch news broadcasts (NOS 8 uur Journaal). It shows how either available collateral data sources (subtitling information for the hearing-impaired) or full-text speech recognition transcripts can be used as linguistic content for the generation of time-coded indexes for searching within news shows. Although the subtitling information in itself would already be enough to enable access, speech recognition transcripts are generated as well for demonstration purposes. The subtitling information is captured using a teletext capturing card and synchronized with the video using a manually determined off-set value. The speech recognition system consists of decision-tree state-clustered acoustic models trained on approximately 20 hours of speech from the Spoken Dutch Corpus [10], a vocabulary of 65K words extracted from a newspaper collection and a 3-gram language model trained on some 300M words of newspaper text data. Currently, the speech recognition system is static; it does not update the vocabulary and language model, nor does it perform any acoustic adaptation schemes. The incorporation of such procedures is scheduled for a new version of the demonstrator.

As the subtitling information provides information on topic boundaries, we can use real topic boundaries for the segmentation of the news show into 'subdocuments'. In case we have to rely on ASR transcripts only, the segmentation news can be based on acoustic information such as speaker changes, speech/non-speech transitions and silences.

In order to demonstrate the added value of links that are automatically generated across media types via high level annotation, the linguistic annotations of the news items (either based on subtitles or ASR) are linked to an up-to-date database of Dutch newspaper articles made available for demonstration purposes by PCM publishers[5]. The links from broadcast news fragments to related newspaper articles are generated by (i) using a stopped version of the textual video annotation as a query for a search in the newspaper archive, (ii) matching the query with the content in the newspaper archive using Okapi term weighting, and (iii) presenting the top-n results in a clickable list, ordered by date or by relevance.

7.2 Novalist News Browser

The broadcast news browser described above primarily demonstrates the added value of automatic linguistic annotation of audiovisual content. The functionality of the so-called Novalist browser (developed at TNO) can be regarded as complementary: it aims to facilitate the work of information analysts in the following way: (i) related news stories are clustered to create dossiers or 'threads', (ii) dossiers resulting from clustering are analyzed and automatically annotated with several types of metadata, and (iii) a browsing screen provides multiple views on the dossiers and their metadata. All analysis steps can be performed data-driven.

The corpus covered by the Novalist demonstrator consists of a collection of news items published by a number of major Dutch newspapers and magazines, web crawls, a video corpus of several news magazines and a video archive with all 2001 news broadcasts of *NOS 8 uur Journaal*. Here, the teleprompter files for the video archive function as collateral text. Transcripts of broadcast audio generated with automatic speech recognition (ASR) can also be incorporated. The entire demonstrator collection consists of some 160,000 individual news items from 21 different sources, and recent new releases even more.

The system has to deal with dynamic information, about which no full prior knowledge is available. There is no fixed number of target topics and events types. The system must both discover new events as the incoming stories are processed, and associate incoming stories with the event-based story clusters already created. Document clustering is done incrementally: for a new incoming story, the system has to decide instantaneously to which topic cluster the story belongs. Since the clustering algorithms are unsupervised, no training data is needed.

Via document clustering, structure is generated in news streams, while the annotations can be applied as filters: search for relevant items can be limited to relevant subsets of the collection. Novalist dossiers are visualized in a compact

---

[5] PCM publishers is one of the largest publishers in the Dutch language region: http://www.pcmuitgevers.nl/

overview window with links to a time axis. Additional functionality consists of the automatic generation of links to related sources, both internal and external. For a detailed explanation of the concept of topic detection and the similarity concept applied in the language modeling approach that is underlying Novalist, and for an overview of the performance evaluation of some components, cf. [14], [5].

8   Conclusion

The two demonstration browsers described here show how automatically extracted annotation based on non-image features can successfully support the exploitation of multimedia content. The possibility to link textual content from diverse sources to media files and vice versa, strengthens the impact of audio and speech analysis. The transcript processing techniques deployed can be linked to query functionality at several levels of conceptual abstraction: from the words spoken to higher level semantic concepts that have automatically detected in the textual content via clustering and classification. Future research will be directed towards the integration with metadata generation based on visual analysis.

Acknowledgment

This work was partly supported by the Dutch bsik-programme MultimediaN (http://www.multimedian.nl) and the EU projects AMI (IST-FP6-506811), MESH (IST-FP6-027685), and MediaCampaign (IST-PF6-027413).

References

1. A. Allauzen and J.L. Gauvain. Diachronic vocabulary adaptation for broadcast news transcription. In *InterSpeech*, Lisbon, September 2005.
2. C. Auzanne, J.S. Garofolo, J.G. Fiscus, and W.M Fisher. Automatic Language Model Adaptation for Spoken Document Retrieval. In *Proceedings of RIAO 2000, Content-Based Multimedia Information Access*, pages 132–141, 2000.
3. M. G. Brown, J.T. Foote, G.J.F. Jones, K. Sparck Jones, and S. J. Young. Automatic Content-based Retrieval of Broadcast News. In *Proceedings of the third ACM international conference on Multimedia*, pages 35–43, San Francisco, November 1995. ACM Press.
4. Lin Chase. Blame assignment for errors made by large vocabulary speech recognizers. In *proceedings Eurospeech '97*, pages 1563–1566, Rhodes, Greece, 1997.
5. F.M.G. de Jong and W. Kraaij. Content reduction for cross-media browsing. In H. Saggion and J.-L. Minel, editors, *RANLP workshop 'Crossing Barriers in Text Summarization Reserach*, pages 64–69, Borovets, Bulgaria, 2005.
6. J.S. Garofolo, C.G.P. Auzanne, and E.M Voorhees. The TREC SDR Track: A Success Story. In *Eighth Text Retrieval Conference*, pages 107–129, Washington, 2000.

7. P. Jourlin, S.E. Johnson, K. Spärck Jones, and P.C. Woodland. General Query Expansion Techniques for Spoken Document Retrieval. In *Proc. ESCA Workshop on Extracting Information from Spoken Audio*, pages 8–13, Cambridge, UK, 1999.
8. W. Kraaij, J. van Gent, R. Ekkelenkamp, and D. van Leeuwen. Phoneme based spoken document retrieval. In *Proceedings of the fourteenth Twente Workshop on Language Technology TWLT-14*, pages 141–153, University of Twente, 1998.
9. Pedro J. Moreno, Chris Joerg, Jean-Manuel Van Thong, and Oren Glickman. A Recursive Algorithm for the Forced Alignment of Very Long Audio Segments. In *Proceedings of the 5th International Conference on Spoken Language Processing (ICSLP'98)*, Sydney, Australia, 1998.
10. N. Oostdijk. The Spoken Dutch Corpus. Overview and first evaluation. In M. Gravilidou, G. Carayannis, S. Markantonatou, S. Piperidis, and G. Stainhaouer, editors, *Second International Conference on Language Resources and Evaluation*, volume II, pages 887–894, 2000.
11. R.J.F. Ordelman. *Dutch Speech Recognition in Multimedia Information Retrieval*. Phd thesis, University of Twente, Enschede, October 2003. publisher: Taaluitgeverij Neslia Paniculata publisherlocation: Enschede, ISSN: 1381-3617; No 03-56, ISBN: 90-75296-08-8, Numberofpages: 268.
12. O. Siohan, T. Myrvol, and C. Lee. Structural maximum a posteriori linear regression for fast hmm adaptation, 2000.
13. A.F Smeaton, W. Kraaij, and P. Over. Trecvid - an overview. In *Proceedings of TRECVID 2003*, USA, 2003. NIST.
14. Martijn Spitters and Wessel Kraaij. Unsupervised clustering in multilingual news streams. In *Proceedings of the LREC 2002 workshop: Event Modelling for Multilingual Document Linking*, pages 42–46, 2002.
15. Khiet P. Truong and David A. van Leeuwen. Automatic detection of laughter. In *InterSpeech*, pages 485–488, Lisbon, September 2005.
16. David van Leeuwen and Marijn Huijbregts. The ami speaker diarization system for nist rt06s meeting data. In *in NIST 2006 Spring Rich Transcrition Evaluation Workshop*, Washington DC, USA, 2006.
17. Thijs Westerveld, Arjen P. de Vries, and Georgina Ramírez. Surface features in video retrieval. In *3rd International Workshop on Adaptive Multimedia Retrieval, AMR'05*, 2005.
18. P.C. Woodland, S.E. Johnson, P. Jourlin, and K. Spärck Jones. Effects of Out of Vocabulary Words in Spoken Document Retrieval. In *2000 ACM SIGIR Conference*, pages 372–374, Athens Greece, 2000.
19. U. Yapanel and J. H. L. Hansen. A new perspective on feature extraction for robust in-vehicle speech recognition. In *Proceedings of Eurospeech*, pages 1281–1284, 2003.

Author Index

Lecture Notes in Computer Science

For information about Vols. 1–4216

please contact your bookseller or Springer